AF772889

DESALINATION
A NATIONAL PERSPECTIVE

Committee on Advancing Desalination Technology

Water Science and Technology Board

Division on Earth and Life Studies

NATIONAL RESEARCH COUNCIL
OF THE NATIONAL ACADEMIES

THE NATIONAL ACADEMIES PRESS
Washington, D.C.
www.nap.edu

THE NATIONAL ACADEMIES PRESS 500 Fifth Street, N.W. Washington, DC 20001

NOTICE: The project that is the subject of this report was approved by the Governing Board of the National Research Council, whose members are drawn from the councils of the National Academy of Sciences, the National Academy of Engineering, and the Institute of Medicine. The members of the panel responsible for the report were chosen for their special competences and with regard for appropriate balance.

Support for this study was provided by the U.S. Bureau of Reclamation under Grant No. 06CS811198. Any opinions, findings, conclusions, or recommendations expressed in this publication are those of the author(s) and do not necessarily reflect the views of the organizations or agencies that provided support for the project.

International Standard Book Number-13: 978-0-309-11923-8 (Book)
International Standard Book Number-10: 0-309-11923-5 (Book)
International Standard Book Number-13: 978-0-309-11924-5 (PDF)
International Standard Book Number-10: 0-309-11924-3 (PDF)
Library of Congress Catalog Card Number: 2008933274

Additional copies of this report are available from the National Academies Press, 500 5th Street, N.W., Lockbox 285, Washington, DC 20055; (800) 624-6242 or (202) 334-3313 (in the Washington metropolitan area); Internet, http://www.nap.edu.

Copyright 2008 by the National Academy of Sciences. All rights reserved.

Printed in the United States of America.

THE NATIONAL ACADEMIES

Advisers to the Nation on Science, Engineering, and Medicine

The **National Academy of Sciences** is a private, nonprofit, self-perpetuating society of distinguished scholars engaged in scientific and engineering research, dedicated to the furtherance of science and technology and to their use for the general welfare. Upon the authority of the charter granted to it by the Congress in 1863, the Academy has a mandate that requires it to advise the federal government on scientific and technical matters. Dr. Ralph J. Cicerone is president of the National Academy of Sciences.

The **National Academy of Engineering** was established in 1964, under the charter of the National Academy of Sciences, as a parallel organization of outstanding engineers. It is autonomous in its administration and in the selection of its members, sharing with the National Academy of Sciences the responsibility for advising the federal government. The National Academy of Engineering also sponsors engineering programs aimed at meeting national needs, encourages education and research, and recognizes the superior achievements of engineers. Dr. Charles M. Vest is president of the National Academy of Engineering.

The **Institute of Medicine** was established in 1970 by the National Academy of Sciences to secure the services of eminent members of appropriate professions in the examination of policy matters pertaining to the health of the public. The Institute acts under the responsibility given to the National Academy of Sciences by its congressional charter to be an adviser to the federal government and, upon its own initiative, to identify issues of medical care, research, and education. Dr. Harvey V. Fineberg is president of the Institute of Medicine.

The **National Research Council** was organized by the National Academy of Sciences in 1916 to associate the broad community of science and technology with the Academy's purposes of furthering knowledge and advising the federal government. Functioning in accordance with general policies determined by the Academy, the Council has become the principal operating agency of both the National Academy of Sciences and the National Academy of Engineering in providing services to the government, the public, and the scientific and engineering communities. The Council is administered jointly by both Academies and the Institute of Medicine. Dr. Ralph J. Cicerone and Dr. Charles M. Vest are chair and vice chair, respectively, of the National Research Council.

www.national-academies.org

COMMITTEE ON ADVANCING DESALINATION TECHNOLOGY

AMY K. ZANDER, *Chair*, Clarkson University, Potsdam, New York
MENACHEM ELIMELECH, Yale University, New Haven, Connecticut
DAVID H. FURUKAWA, Separation Consultants Inc., Poway, California
PETER GLEICK, Pacific Institute for Studies in Development, Environment, and Security, Oakland, California
KENNETH R. HERD, Southwest Florida Water Management District, Brooksville, Florida
KIMBERLY L. JONES, Howard University, Washington, DC
PHILIP ROLCHIGO, Pentair, Inc., Golden Valley, Minnesota
SANDEEP SETHI, Carollo Engineers, Sarasota, Florida
JOHN TONNER, Water Consultants International, Mequon, Wisconsin
HENRY J. VAUX, University of California, Berkeley
JUDITH S. WEIS, Rutgers University, Newark, New Jersey
WARREN W. WOOD, Michigan State University, East Lansing

NRC Staff

STEPHANIE E. JOHNSON, Study Director
LAURA J. EHLERS, Interim Study Director (October 2006 to January 2007)
MICHAEL J. STOEVER, Senior Program Assistant

Preface

Water is necessary for life. Provision of a safe, reliable, and sustainable water supply to communities is an essential function of water utilities in the United States. As climate changes, population grows, and local water scarcity concerns heighten, desalination of brackish water and seawater is increasingly considered as an option for a source of new water to meet anticipated water supply needs. Desalination opens the door to conversion of the vast ocean and brackish inland water bodies into usable water for municipalities. Given the possibilities for clean water supply, desalination technology adoption in the United States lags behind that of many other countries. Concerns have been raised that technological barriers and financial costs prohibit broader desalination implementation.

The committee formed by the Water Science and Technology Board of the National Research Council (NRC) performed a critical analysis of current desalination technologies and the barriers to widespread implementation and addressed the development of a national strategic research agenda for desalination. The report presents a brief history of desalination research and research funding in the United States, presents issues of water sufficiency and the potential for desalination to meet anticipated supply needs, and outlines the current state of the science in desalination technology. Environmental issues are examined, along with the costs of the technology and recent trends in cost compared with other water supply options, including conservation. Practical implementation aspects are analyzed. The issues are combined, leading to a framework for a strategic national research agenda in desalination, involving both federal and nonfederal inputs. The agenda is needed to understand and mitigate environmental impacts of source water withdrawal and concentrate discharge and to marginally reduce the financial costs of desalination.

I thank the committee members for the contribution of their unique and individual expertise, intellect, and time to the preparation of this consensus report. The committee expertise comes from many areas, all of which were essential to the preparation of the report under a rather aggressive time schedule. The committee members offered many hours of personal time to the preparation of a report of which they can all be proud. Each member contributed consistently and tirelessly to the overall report. Committee views were all considered carefully in the study process, and I thank them for their technical and intellectual contributions and collaborative spirit.

The successful preparation of this report was also in large part due to the NRC staff members for their skills and dedication. Stephanie Johnson, senior staff officer, made certain committee contributions were timely and on task, contributed personally to writing sections of the report, and thoroughly edited the completed report to bring the various committee contributions to a single voice. Her persistence, attention to detail, and organization were essential to the quality of the final product. The committee was quite capably assisted by Michael Stoever, project assistant, who handled administrative details of the meetings and aided in report preparation. Thanks are also due to Laura Ehlers, senior staff officer, who stepped in to take the reins for a few months early in the committee process when her assistance was needed and greatly appreciated.

The committee is grateful to the sponsors who provided support for the study. This report was undertaken at the request of the Department of the Interior—Bureau of Reclamation and the U.S. Environmental Protection Agency. The sponsors gave generously of their time as well, informing the committee of issues they face and responding to questions from the group.

This report was reviewed in draft form by individuals chosen for their breadth of perspectives and technical expertise in accordance with the procedures approved by the National Academies' Report Review Committee. The purpose of this independent review was to provide candid and critical comments to assist the institution in ensuring that its published report is scientifically credible and that it meets institutional standards for objectivity, evidence, and responsiveness to the study charge. The reviewer comments and draft manuscript remain confidential to protect the deliberative process. We thank the following reviewers for their helpful suggestions, all of which were considered and many of which were wholly or partly incorporated in the final report: Robert Cheng, Long Beach Water Department; Lisa Henthorne-Jenkel, CH2M Hill, Inc.; Sabine Lattemann, University of Oldenburg; Noam Lior, University of Pennsylvania; Daniel P. Loucks, Cornell University; Thomas Luster,

California Coastal Commission; Michael Mickley, Mickley & Associates; Jonas Minton, Planning and Conservation League; Hayden H. Murray, University of Indiana; Robert L. Riley, Separation Systems Technology, Inc.; David Stewart, Stewart Environmental Consultants, Inc.; R. Rhodes Trussell, Trussell Technologies, Inc.; Dale Whittington, University of North Carolina; and Alicia Wilson, University of South Carolina.

Although these reviewers provided many constructive comments and suggestions, they were not asked to endorse the conclusions and recommendations nor did they see the final draft of the report before its release. The review of this report was overseen by Michael C. Kavanaugh, Malcolm Pirnie, Inc., and Marcia K. McNutt, Monterey Bay Aquarium Research Institute. Appointed by the NRC, they were responsible for making certain that an independent examination of this report was carried out in accordance with institutional procedures and that all review comments received full consideration. Responsibility for the final content of this report rests entirely with the authoring committee and the NRC.

Amy K. Zander, *Chair*
Committee on Advancing Desalination Technology

Acknowledgments

Many individuals assisted the committee and the National Research Council staff in their task to create this report. We would like to express our appreciation to the following people who have provided presentations to the committee, served as guides during the field trips, and provided comments to the committee:

Presentations

Dave Akers, Colorado Water Quality Protection Division
Ashie Akpoji, South Florida Water Management District
Rick Arber, Arber and Associates
Edmund Archuleta, El Paso Desalination Facility
Paul Armistead, Office of Naval Research
Jorge Arroyo, Texas Water Development Board
Richard Atwater, Desalination in the Inland Empire
Maryanne Bach, U.S. Bureau of Reclamation
Mark Barcelo, Southwest Florida Watershed Management District
Robert Bastian, U.S. Environmental Protection Agency
Richard Bell, Dana Point Ocean Desalination Project
Ann Codrington, U.S. Environmental Protection Agency
Jeff Graham, Scripps Institute
Bruce Muller, U.S. Bureau of Reclamation
Robert Hirsch, U.S. Geological Survey
Scott Jenkins, Scripps Institute
Fawzi Karajeh, California Office of Water Use Efficiency and Transfers
Richard Kottenstette, Sandia National Laboratories
Jim Kundell, Carl Vinson Institute of Government UGA
Chris Rayburn, AWAARF
Robert Reiss, Reiss Environmental
Michelle K. Robinson, APR
Dave Stewart, Stewart Environmental Consultants, Inc.
Barbara Vergara, St. John's River Watershed Management District
Robert Yamada, San Diego Water Authority

Field Trip Guides

Robert Cheng, Long Beach Water Department
Ken Herd, Tampa Bay Water
Eric Sabolsice, American Water-Pridesa
Kevin Wattier, Long Beach Water Department

Contents

Summary

Efforts to identify new, untapped sources of supply have dominated water policy for the past century. There has been an exponential increase in desalination capacity both globally and nationally since 1960, fueled in part by this growing concern for local water scarcity and made possible to a great extent by a major federal investment for desalination research and development in the late 1950s to the early 1980s. Traditional sources of supply are increasingly expensive, unavailable, or controversial, but desalination technology offers the potential to substantially reduce water scarcity by converting the almost inexhaustible supply of seawater and the apparently vast quantities of brackish groundwater into new sources of freshwater. Although total water use in the United States has remained steady over recent decades, interest in brackish water and seawater desalination will likely continue, particularly in water-scarce regions, in localities experiencing rapid population growth, or where users are able and willing to pay for a high-quality, reliable new supply.

Historically, the high cost and energy requirements of desalination had confined its use to places where energy is inexpensive and freshwater scarce. Recent advances in technology, especially improvements in membranes, have made desalination a realistic water supply option. The cost of desalinating seawater in the United States is now competitive with other alternatives in some locations and for some high-valued uses, and there is considerable interest in hastening the time when costs of desalination are routinely competitive with the costs of alternatives.

With support from the U.S. Bureau of Reclamation and the U.S. Environmental Protection Agency, the National Research Council convened the Committee on Advancing Desalination Technology to assess the state of the art in relevant desalination technologies and factors such as cost and implementation challenges. The committee also was asked to describe reasonable long-term goals for advancing desalination technology and to provide recommendations for action and research. Finally, the committee was asked to estimate the funding necessary to support the proposed research agenda and to identify appropriate roles for governmental and nongovernmental entities. This report builds on a 2004 National Research Council report that provided a scientific assessment of the Bureau of Reclamation's and Sandia National Laboratories' *Desalination and Water Purification Technology Roadmap,* or *Roadmap*, which was intended to serve as a strategic pathway for future desalination and water purification research (USBR and Sandia National Laboratories, 2003).

WATER FOR THE FUTURE: THE ROLE OF DESALINATION

When considering future water supplies, it is important to recognize that past patterns of water use will not always be a reliable indicator of future demand. In particular, the assumption that water demands will inevitably parallel population and economic growth no longer appears to be correct. Nevertheless, water scarcity in some regions of the United States will certainly intensify over the coming decades, and no one option or set of options is likely to be sufficient to manage this intensifying scarcity. Desalination, using both brackish and seawater sources, is likely to have a niche in the future water management portfolio of the United States.

The committee was specifically asked to address the potential for seawater and brackish water desalination to help meet anticipated water supply needs in the United States. The committee concluded that the potential for desalination cannot be definitively determined because it depends on a host of complicated and locally variable economic, social, environmental, and political factors. In the complete absence of these factors, the theoretical potential for desalination is effectively unlimited. Large quantities of inland brackish groundwater appear to be available for development; in coastal areas, ocean resources are essentially infinite in comparison to human demands. But, as with most resource questions, the theoretical potential and the practical potential are far different. All water management and planning takes place in the context of economic, social, environmental, and political factors, and these factors are far more important than technological desalination process constraints in limiting the potential for desalination to help meet anticipated water supply needs. As a result, this report addresses key technological issues that may lend themselves to focused research and development efforts, but the report also addresses nontechnical questions that may ultimately prove to be more limiting.

The costs of producing desalinated water—the cost of removing salts to create freshwater—is no longer the primary barrier to implementing desalination technology, because there have been significant reductions in desalination production costs. Meanwhile, the costs of other alternatives for augmenting water supplies have continued to rise, often making desalination more attractive in a relative sense. A continuation of these trends would likely make desalination costs more attractive and less of a constraint in the future. Nevertheless, concentrate management costs vary widely, depending on local regulations and site-specific conditions, and have generally increased in recent decades. Where low-cost concentrate management alternatives are not available, the financial costs of desalination can be prohibitive. There is also considerable uncertainty about the environmental impacts of desalination and, consequently, concern over its potential effects. Possible environmental impacts of desalination are impingement and entrainment of organisms when seawater is taken in, ecological impacts from disposing of salt concentrates, and increased greenhouse gas emissions from increased energy use, among other concerns. Although limited studies to date suggest that the environmental impacts *may* be less detrimental than many other types of water supply, site-specific information necessary to make detailed conclusions on environmental impacts is typically lacking.

A strategic desalination research and development effort can help make desalination a more attractive water supply option for communities facing water shortages and can enable desalination technology to serve a larger role in addressing the nation's water demands. The two main goals of this research and development effort should be (1) to understand the environmental impacts of desalination and develop approaches to minimize these impacts relative to other water supply alternatives, and (2) to develop approaches to lower the financial costs of desalination so that it is an attractive option relative to other alternatives in locations where traditional sources of water are inadequate. Success of the proposed research agenda will depend on coordinated federal leadership and participation by state and local governments, nongovernmental organizations, and the private sector.

CURRENT INVESTMENT IN DESALINATION RESEARCH AND DEVELOPMENT

Based on the committee's survey of federal investment in desalination research and development together with estimates and analyses of state, nongovernmental organizations, and private-sector funding, the committee reached several conclusions (see also Chapter 2) that underscore the need for the development of a national strategic research agenda for desalination:

- **There is no integrated and strategic direction to the federal desalination research and development efforts.** Desalination research and development efforts are funded through at least nine federal agencies and laboratories, each with their own research objectives and priorities. The majority of federal desalination research and development funding also comes from congressional earmarks, which limits the ability to develop a steady research program. Federal funding for desalination research declined from approximately $24 million per year in fiscal years (FYs) 2005 and 2006 to $10 million in FY 2007 (a decline of nearly 60 percent), largely due to an absence of earmarks in FY 2007.
- **Some states, especially California, have also made sizeable recent investments in desalination research and development.** The majority of this funding is being directed toward site- or region-specific problems, with heavy emphasis on pilot and demonstration projects.
- **The private sector appears to fund the majority of desalination research, with total annual spending estimated to be more than twice that of all other surveyed sources of such funding.** Based on the judgment of individuals working in the industry, the private sector allocates a smaller fraction of its research portfolio to high-risk desalination research than the federal government. Given the large research investment, however, private-sector funding for high-risk activities is estimated to be roughly equivalent to federal funding for high-risk research.

STATE OF DESALINATION TECHNOLOGY

The state of desalination technology, including intakes, pretreatment, desalination processes, post-treatment, and concentrate management, is outlined in Chapter 4. Industry has made great strides in reducing energy use for desalination with the commercialization of high-efficiency energy recovery devices and improvements in membrane technology. Reverse osmosis (RO) technology is relatively mature, and current energy use is within a factor of 2 of the theoretical thermodynamic minimum value for separating solutes from water. Thus, RO is the standard by which novel desalination technologies should be assessed when specific energy use is the main consideration. Meanwhile, opportunities exist to further reduce cost and energy use of current technologies by small but economically significant amounts:

- **In RO desalination, the costs and energy requirements of water production can be further reduced by mitigating fouling through pretreatment; developing high-permeability, fouling-resistant, high-rejection, oxidant-resistant membranes; and optimizing membrane**

module and membrane system design. Operating the RO process at lower hydraulic pressure while maintaining high throughput is the key to reducing the specific energy for membrane-based desalination. To fully utilize the capacity of high-permeability RO membranes and to accommodate even more permeable RO membranes in the future, it is imperative to reduce fouling and concentration polarization effects and to develop new module configurations and system designs to avoid or overcome thermodynamic restriction. Fouling can be reduced by a more robust pretreatment and by the development of fouling-resistant membranes. Practical and economic constraints, however, are likely to inhibit RO energy use from decreasing more than approximately 15 percent below the values of the best current technology. This level of improvement would still be valuable for reducing cost and energy use, but greater returns on investment than this should not be expected.

- **Seawater desalination using thermal processes can be cost effective when waste heat is utilized effectively.** Location of low-grade and/or waste heat resources near large water consumers may reduce the cost of heat energy and offset the higher specific-energy requirements of thermal desalination when compared to RO. Hybrid membrane–thermal desalination approaches offer additional operational flexibility and opportunities for water production cost savings for facilities co-located with power plants.
- **Few, if any, cost-effective environmentally sustainable concentrate management options exist for inland desalination facilities.** Several methods are currently available for concentrate management, and each method has its own set of site-specific costs, benefits, regulatory requirements, environmental impacts, and limitations. Low- to moderate-cost inland disposal options can be limited by the salinity of the concentrate and by location and climate factors. Only evaporation ponds and high-recovery/thermal evaporation systems are zero-liquid discharge solutions, but high costs limit their consideration for most municipal applications.

ENVIRONMENTAL IMPACTS

Knowledge of the potential environmental impacts of desalination processes is essential to water supply planners when they consider desalination among many water supply alternatives. All components of the water use cycle should be considered, including source water impacts, the likely greenhouse gas emissions from the energy requirements of the desalination process, potential impacts from concentrate management approaches, and environmental health considerations in the product water. Ideally, these considerations should be compared against equally rigorous environmental impact analyses of water supply alternatives, so that decisions can be made based on comparisons of the full economic costs and benefits, including environmental and social costs and benefits, among the various water supply alternatives.

There is a considerable amount of uncertainty about the environmental impacts of desalination and, consequently, concern over its potential effects. Therefore, the following research is recommended:

- **Site-specific assessments of the impacts of source water withdrawals and concentrate management should be conducted and the results synthesized in a national assessment of potential impacts.** The ecological effects of concentrate discharge into the ocean appear to vary widely and depend on the site-specific environment, the organisms examined, the amount of dilution of the concentrate, and the use of diffuser technology. The ecological impacts of surface water intakes have been well studied for power plants but not for desalination plants, and these impacts will likely vary from place to place. General information on potential impacts from groundwater withdrawal and injection are available from decades of hydrogeologic studies for other purposes, but site-specific analyses are necessary to understand the impacts from a proposed facility. Once a number of rigorous site-specific studies are conducted, this information should be synthesized, using other existing data, to develop an overarching assessment of the possible range of impacts from desalination in the United States.
- **Monitoring and assessment protocols should be developed for evaluating the potential ecological impacts of surface water concentrate discharge.** Adequate site-specific studies on potential biological or ecological effects are necessary prior to the development of desalination facilities, and planners would benefit from clear guidance on appropriate monitoring and assessment protocols. Specific recommendations are provided in Chapter 5.
- **Longer-term, laboratory-based assays on the sublethal effects of concentrate discharge should be conducted to understand the range of environmental impacts from desalination plants.** Except for a few short-term lethality studies that do not give insight into long-term effects, little research has been done on the impacts of concentrate discharges on organisms in receiving waters. Longer-term laboratory-based biological assays should evaluate impacts of concentrate on development, growth, and reproduction using a variety of different organisms, including those native to areas where desalination plants are proposed. These results can be utilized in a risk assessment framework.

- **Water quality guidance, based on an analysis of the human health effects of boron in drinking water and considering other sources of exposure, is needed to support decisions for desalination process design.** There are concerns about boron in product water from seawater desalination because the boron levels after single-pass RO commonly exceed current WHO health guidelines and the EPA health reference level. A range of water quality levels (0.5 to 1.4 mg/L) have been proposed as protective of public health based on different assumptions in the calculations. The EPA has decided not to develop a maximum contaminant level for boron because of its lack of occurrence in most groundwater and surface water and has encouraged affected states to issue guidance or regulations as appropriate. Therefore, most U.S. utilities lack clear guidance on what boron levels in drinking water are suitably protective of public health. Boron can be removed through treatment optimization, but that treatment could aversely affect the cost of seawater desalination.
- **Further research and applications of technology should be carried out on how to mitigate environmental impacts of desalination and reduce potential risks relative to other water supply alternatives.** For example, intake and outfall structures could be designed to minimize impingement and entrainment and to encourage improved dispersion of the concentrate in coastal discharges. Research could also explore beneficial reuse of the desalination by-products and develop technologies that reduce the volume of this discharge.

Desalination efforts do not need to be halted until this research is done and uncertainties are removed, but research investments should be made to help reduce potential risks.

COSTS AND BENEFITS OF DESALINATION

Historically, the relatively high financial costs of water production via desalination have constrained the use of desalination technologies in all but a few very specific circumstances. As discussed earlier, the financial cost picture has changed in a number of important ways. There have been significant reductions in membrane costs and improvements in the energy efficiency of the desalination process. Perhaps more significant, the costs of other alternatives for augmenting water supplies have continued to rise, making desalination production costs more attractive in a relative sense. The trend of cost reduction may be abetted through a program of strategically directed research aimed at achieving potentially large cost reductions. Nevertheless, the costs of concentrate management are potentially large and vary from site to site. Such costs have the potential to offset reductions in water production costs. The following conclusions about the cost and benefits of desalination are based on the discussion and analyses in Chapter 6:

- **Substantial reductions in the financial cost of producing desalinated water will require substantial reductions in either energy costs or capital costs**. Energy and capital costs are the two largest components of financial cost for both thermal and membrane seawater desalination processes. Future trends in energy costs also will be important inasmuch as significant increases in energy prices could offset or more than offset cost reductions in other areas and make desalination technologies less attractive. Cost savings are possible if novel technologies or configurations are developed that optimize the use of alternative energy sources, including low-grade or waste energy.
- **For brackish water desalination, the costs of concentrate management can vary enormously from project to project and may rival energy and capital costs as the largest single component of cost.** The high cost of environmentally sustainable concentrate management at some inland locations ultimately offsets the cost advantage that can be obtained from utilizing feed waters with lower salinity.
- **Conservation and transfers from low- to high-valued uses will usually be less costly than supply augmentation schemes, including desalination.** In many circumstances, there remain methods of demand management that can make significant additional quantities of water available at less cost than desalination. Similarly, market-like transfers of water can also offer relatively low-cost ways of acquiring additional supplies of water. Conservation and efficiency improvements that reduce the total demand for water often come with associated benefits, such as reduced energy costs.
- **There are small but significant efficiencies that can be made in membrane technologies that will reduce the energy needed to desalinate water and, therefore, offer potentially important process cost reductions.** Development of membranes that operate effectively at lower pressures could lead to 5 to 10 percent reductions in total costs of the desalination process associated with a 15 percent decrease in energy use. In contrast, extending membrane life beyond the current 5-year design life is likely to have a small impact on desalination costs because membranes account for a minimal proportion of total costs. Prevention of catastrophic failure through robust pretreatment is important because membrane failure within the first year of operation can cause an annual cost increase of more than 25 percent.

- **To make the true costs of desalination transparent, the economic costs should be accounted for and reported accurately.** Failure to price water accurately can lead to inefficient use and overuse. Average cost pricing understates the cost of desalinated water to the consumer, and the supplier should take care in reporting the true and accurate economic costs publicly.

A STRATEGIC DESALINATION RESEARCH AGENDA

Over the past 50 years the state of desalination technology has advanced substantially. Even so, concerns about potential environmental impacts continue to limit the application of desalination technology in the United States, and desalination remains a higher-cost alternative for water supply in many communities. In order for desalination to become a more attractive water supply option for communities facing water shortages, two overarching long-term research goals need to be met:

1. Understand the environmental impacts of desalination and develop approaches to minimize these impacts relative to other water supply alternatives, and

2. Develop approaches to lower the financial costs of desalination so that it is an attractive option relative to other alternatives in locations where traditional sources of water are inadequate.

Although the environmental impacts of coastal desalination may be less than that of other water supply alternatives, the uncertainty about potential site-specific impacts and their mitigation for both coastal and inland operations are large barriers to the application of desalination in the United States. This uncertainty leads to stakeholder disagreements and a lengthy and costly planning and permitting process. Without rigorous scientific research to identify specific potential environmental impacts (or a lack of impacts), planners cannot assess the feasibility of desalination at a site or determine what additional mitigation steps are needed.

At present, desalination costs are already low enough to make desalination an attractive option for some communities when the benefits of desalination are considered, such as providing a drought-resistant supply and providing a means to diversify a large community's water supply portfolio. However, the costs of desalination, like the costs of water supply alternatives, are locally variable and are influenced by factors such as site conditions and concentrate management options. In addition, increasing awareness of potential environmental impacts is raising the costs of permitting and intake and outfall configurations in the United States. Meanwhile, the future costs of energy are uncertain.

A strategic research agenda in support of these two overarching goals is described in Chapter 8 (see Box S-1). This research agenda is broadly conceived and includes research that could be appropriately funded and conducted in either the public or the private sector. Several recommendations for implementing the proposed research agenda follow:

- **A coordinated, strategic plan should be developed to ensure that future federal investments in desalination are integrated and prioritized and address the two major goals identified in this report.** The strategic application of federal funding for desalination research can advance the implementation of desalination technologies in areas where traditional sources of water are inadequate. Responsibility for developing the plan should rest with the Office of Science and Technology Policy (OSTP). Initial federal appropriations on the order of recent spending on desalination research (total appropriations of about $25 million annually) should be sufficient to make good progress toward these goals, when complemented by ongoing nonfederal and private-sector desalination research, if the funding is directed toward the proposed research topics recommended in this report. Reallocation of current federal spending will be necessary to address currently underfunded topics. If current federal research and development funding is not reallocated, new appropriations will be necessary. However, support for the research agenda stated here should not come at the expense of other high-priority water resources research topics. Five years into the implementation of this plan, the OSTP should evaluate the status of the plan, whether goals have been met, and the need for further funding.
- **Environmental research should be emphasized up front when implementing the research agenda.** Uncertainties regarding environmental impacts and ways to mitigate these impacts are some of the largest hurdles to implementation of desalination in the United States, and research in these areas has the greatest potential for enabling desalination to help meet future water needs in communities facing water shortages. Priority areas of environmental research are discussed in Chapters 5 and 8.
- **Research funding in support of reducing the costs of desalination should be directed strategically toward research topics that are likely to make improvements against benchmarks set by the best current technologies for desalination.** Because the private sector is already making impressive strides toward reducing the process costs of desalination, the federal research funding should emphasize long-term,

BOX S-1
Priority Research Areas

The committee has identified priority research areas to help make desalination a competitive option among water supply alternatives for communities facing water shortages. These research areas, which are described in more detail in Chapter 8, are summarized here. The highest priority topics are shown in bold. Some of this research may be most appropriately supported by the private sector. The research topics for which the federal government should have an interest—where the benefits are widespread and where no private-sector entities are willing to make the investments and assume the risk—are marked with asterisks.

GOAL 1. Understand the environmental impacts of desalination and develop approaches to minimize these impacts relative to other water supply alternatives

1. **Assess environmental impacts of desalination intake and concentrate management approaches****
   a. Conduct field studies to assess environmental impacts of seawater intakes**
   b. Conduct field studies to assess environmental impacts of brackish groundwater development**
   c. Develop protocols and conduct field studies to assess the impacts of concentrate management approaches in inland and coastal settings**
   d. Develop laboratory protocols for long-term toxicity testing of whole effluent to assess long-term impacts of concentrate on aquatic life**
   e. Assess the environmental fate and bioaccumulation potential of desalination-related contaminants**
2. **Develop improved intake methods at coastal facilities to minimize impingement of larger organisms and entrainment of smaller ones****
3. Assess the quantity and distribution of brackish water resources nationwide**
4. Analyze the human health impacts of boron, considering other sources of boron exposure, to expedite water-quality guidance for desalination process design**

GOAL 2. Develop approaches to lower the financial costs of desalination so that it is an attractive option relative to other alternatives in locations where traditional sources of water are inadequate

5. Improve pretreatment for membrane desalination
   a. Develop more robust, cost-effective pretreatment processes
   b. Reduce chemical requirements for pretreatment
6. Improve membrane system performance
   a. Develop high-permeability, fouling-resistant, high-rejection, oxidant-resistant membranes
   b. Optimize membrane system design
   c. Develop lower-cost, corrosion-resistant materials of construction
   d. Develop ion-selective processes for brackish water
   e. Develop hybrid desalination processes to increase recovery
7. Improve existing desalination approaches to reduce primary energy use
   a. Develop improved energy recovery technologies and techniques for desalination
   b. Research configurations and applications for desalination to utilize waste heat**
   c. Understand the impact of energy pricing on desalination technology over time**
   d. Investigate approaches for integrating renewable energy with desalination**
8. Develop novel approaches and/or processes to desalinate water in a way that reduces primary energy use**

GOAL 1 and 2 Cross Cuts.

9. **Develop cost-effective approaches for concentrate management that minimize potential environmental impacts****

high-risk research that may not be attempted by the private sector *and* that is in the public interest. Investigator-driven research should be permitted throughout the proposal process.

Research cannot address all barriers to increased application of desalination technology in regions facing water scarcity concerns; thus, practical desalination implementation issues that pertain to water providers are discussed in Chapter 7. Recommendations in Chapter 7 include building trust and educating the public on desalination project planning, anticipating the sometimes cumbersome regulatory and permitting process, and utilizing pilot testing to optimize the process design.

PROSPECTS FOR DESALINATION

The potential for desalination to meet anticipated water demands in the United States is constrained not by the source water resources or the capabilities of current technology, but by a variety of financial, social, and environmental factors. Substantial uncertainty remains about the environmental impacts of desalination, and resolving these uncertainties and developing methods to mitigate the impacts is the highest priority for future research. Research and development also are needed to continue current trends in reducing the process costs of desalination and developing cost-effective, environmentally sustainable approaches for concentrate management. Implementing the proposed research agenda will require federal leadership and a coordinated, strategic plan among multiple agencies. In addition, the success of the research agenda will depend on participation by federal, state, local agencies, nongovernmental entities, and the private sector.

1

Introduction

Growing concern over local water scarcity and challenges in meeting future water demand has led to heightened interest in desalination technology. In early 2003, the U.S. Bureau of Reclamation and Sandia National Laboratories completed a technology planning activity (the *Desalination and Water Purification Technology Roadmap* or *Roadmap*) intended to serve as a strategic pathway for future desalination and water purification research (USBR and Sandia National Laboratories, 2003). In the fall of 2002, at the request of the Bureau of Reclamation, the National Research Council's (NRC's) Water Science and Technology Board initiated an independent assessment of the Roadmap (NRC, 2004b). NRC (2004b) concluded that in order for desalination technologies to provide safe, reliable, sustainable, and cost-effective water supply for water utilities in the United States, current and anticipated challenges need to be identified and a national research agenda developed. However, the report noted that additional work was needed to build upon the Roadmap and provide thorough, critical analyses of current technologies and research objectives to develop a strategic research agenda for desalination. This report seeks to address these objectives.

SALINE WATER AS A WATER SUPPLY ALTERNATIVE

The Earth contains a vast amount of water, but much of it is too salty for human use without advanced treatment. Nearly all of the Earth's water is found in the world's oceans, while only about 2.5 percent exists as freshwater (see Table 1-1). Much of the freshwater is bound as glaciers and permanent snow, leaving only a small fraction of useable freshwater to meet the world's human water demands and to satisfy environmental needs. As some of the demands for freshwater continue to grow, the availability of new supplies from traditional freshwater sources continues to decline. Therefore, communities are increasingly looking toward more saline waters, such as brackish groundwater or seawater, or otherwise "impaired" waters to address water supply needs.

There are many ways to define the salinity (salt concentration) ranges for fresh and saline waters. Water with greater than 2,000 to 3,000 mg/L total dissolved solids (TDS) is considered too salty to drink (Freeze and Cherry, 1979) or to grow most crops. The World Health Organization considers water with TDS concentrations below 1,000 mg/L to be generally acceptable to consumers, although it notes that acceptability may vary according to circumstances (WHO, 2003). The U.S. Environmental Protection Agency (EPA) notes that drinking water with TDS greater than 500 mg/L can be distasteful (USEPA, 1979). Brackish water has a salinity between that of fresh- and seawater. In more than 97 percent of seawater in the world the salinity is between 33,000 and 37,000 mg/L (Stumm and Morgan, 1996), although the Persian Gulf has an average TDS of 48,000 mg/L (Pankratz and Tonner, 2003). Water with salinity greater than that of seawater is called brine (USGS, 2003).

As noted in Table 1-1, nearly 1 percent of the world's water exists as brackish or saline groundwater. In most inland cases, groundwater salinity results from the dissolution of minerals present in the subsurface, possibly concentrated further by evapotranspiration. Coastal aquifers form another class of brackish water, which is created from the natural mixing of seawater with groundwater that is discharging to the ocean (see also Chapter 5). The thickness of this brackish mixing zone is sometimes increased by coastal groundwater pumping. Brackish groundwater exists at elevations less than 305 m (1,000 feet) across much of the conterminous United States (Feth, 1965) (Figure 1-1) and almost certainly at

TABLE 1-1 Major Stocks of Water on Earth

Location	Amount (10^6 km^3)	Percentage of World Water
Ocean	1338.0	96.5
Glaciers and permanent snow	24.1	1.74
Groundwater (brackish or saline)	12.9	0.94
Groundwater (fresh)	10.5	0.76
Ground ice/permafrost	0.30	0.022
Freshwater lakes	0.091	0.007
Freshwater stream channels	0.002	0.0002

SOURCE: Shiklomanov, 1993.

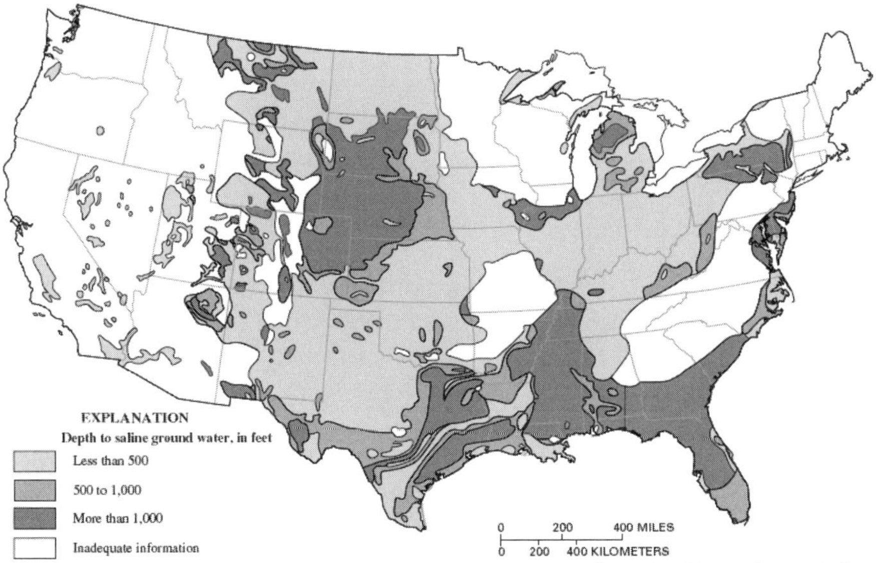

FIGURE 1-1. Depth to brackish groundwater (greater than 1,000 mg/L total dissolved solids) in the conterminous United States.
SOURCE: Generalized from Feth (1965).

comparable depths in Hawaii and Alaska. Both coastal and inland communities are increasingly considering brackish groundwater as a possible water supply resource.

Desalination processes generally treat seawater and brackish waters to produce freshwater (i.e., the desired product stream) and a separate saltier concentrate stream. Several approaches can be used to desalinate saline water sources at the municipal scale. The earliest commercial plants used mostly large-scale thermal evaporation or distillation of seawater. Major facilities were first built in the Persian Gulf region, where excess or inexpensive energy was available and where natural sources of freshwater are relatively scarce. Beginning in the 1970s, plants were installed that used pumps and membranes to produce freshwater, applying the natural biological process of osmosis in reverse. Significant advances in reverse osmosis technology have been achieved in recent years that have reduced the water production costs of desalination. Worldwide, the online capacity[1] for desalination now exceeds 37 million cubic meters of water per day (30,000 acre-feet per day or 10,000 million gallons per

[1] In this report, online capacity includes desalination plants that have been confirmed by Global Water Intelligence (GWI, 2006b) to be online and those that are "presumed online." These online capacity totals do not include plants that were confirmed to be offline, under construction, decommissioned, or "mothballed" or those that were presumed by GWI to be offline.

day) (GWI, 2006b), although this sum represents only about 0.3 percent of total freshwater use (Cooley et al., 2006). More detail on specific processes and technologies is provided in Chapter 4.

STATEMENT OF COMMITTEE TASK AND REPORT OVERVIEW

In 2006, the NRC's Committee on Advancing Desalination Technology was formed to assess the status of desalination technologies and factors such as cost and implementation challenges, and to provide recommendations for action and research. This study was sponsored by the U.S. Bureau of Reclamation and the EPA. The committee was specifically tasked to address the following questions (with cross references to the chapters where the tasks are addressed):

1. **Contributing to the nation's water supplies.** What is the potential for both seawater and inland brackish water desalination to help meet anticipated water supply needs in the United States? (See Chapters 3 and 5.) How do the costs and benefits of desalination compare with other alternatives, including nontechnical options such as water conservation or market transfers of water? (See Chapter 6.)

2. **Assessing the state of technology and setting goals.** What is the current state of the science in desalination technology? (See Chapter 4.) What have the recent trends been (both for seawater and for brackish water) in terms of total cost per unit of water produced and also in the energy efficiency of the process? (See Chapters 4 and 6.) Are there theoretical limits to the efficiency of existing technologies and is there good reason to think that significant advancement can be made toward reaching those limits? (See Chapter 4.) What are reasonable long-term goals for advancing desalination technology? (See Chapter 8.)

3. **Research strategy.** Following up on a recommendation by NRC (2004b) calling for the development of a national research agenda, what research is needed to reach the long-term goals for advancing desalination technology? (See Chapter 8.) What technical barriers should be resolved with existing desalination technologies (including concentrate disposal) and what innovative technologies should be considered? (See Chapters 4 and 5.) In the long-term research agenda for desalination, what balance should be crafted between high-risk research in novel technologies and research that could yield incremental improvements in current technologies? (See Chapter 8.)

4. **Practical aspects of implementation.** What important issues related to implementation must be addressed to significantly improve the applicability of technology for desalination to help meet the nation's water needs (e.g., economics, financing, regulatory, institutional, public acceptance)? (See Chapters 6 and 7.) What are the true economic costs? (See Chapters 5 and 6.) What factors are likely to affect the availability of financing? What are the likely regulatory issues and how easy or difficult will it be to deal with them? Are there other institutional issues? What problems, if any, may arise in ensuring public acceptability of desalination technologies? (See Chapter 7.)

5. **Resources and roles.** What order of magnitude of research funding is needed to significantly advance the field of desalination technology and what are appropriate roles for governmental and nongovernmental entities? (See Chapter 8.)

The committee's conclusions and recommendations are based on a review of relevant technical literature, briefings, and discussions at its six meetings, field trips to desalination facilities (see Acknowledgments), and the experience and knowledge of the committee members in their fields of expertise.

Following this brief introduction, the statement of task is addressed in seven subsequent chapters of this report:

- Chapter 2 provides context for this report by describing the use of desalination technologies in the United States and globally and discussing major research programs—both historical and current—focused on advancing desalination technologies.
- Chapter 3 discusses the issues of water use and water sufficiency and addresses the potential for desalination technologies to help meet anticipated water supply needs.
- In Chapter 4 the state of the science in desalination technology, including intakes, energy recovery, and concentrate management, is described. Current process and technology constraints are discussed, along with the most promising opportunities to maximize energy efficiency, considering the thermodynamic limitations.
- In Chapter 5, environmental issues associated with desalination are discussed, focusing on source water acquisition, concentrate management, human health issues, and potential climate and energy concerns.
- In Chapter 6, the financial and economic circumstances surrounding desalination technology are discussed and the benefits of desalination are examined. The costs of desalination are analyzed to high-

light their major components and the largest opportunities for cost reductions. The chapter also includes a discussion of the costs of desalination relative to other water supply alternatives.

- The practical aspects of implementation for water providers are described in Chapter 7, including regulatory concerns, public perception, and financing.
- In Chapter 8, the committee presents two long-term goals for advancing desalination technology and develops a national research agenda to address these goals. Recommendations are offered on the implementation of this proposed research agenda, including an estimate of the federal resources necessary to support it.

2

Historical and Contemporary Context for Desalination

Humankind has long used basic desalination processes to create drinking water, but advances in research and development over the past 40 years have led to large increases in the use of desalination worldwide. This chapter describes the status of desalination use in the United States and globally. The history of support for research and development that led to today's technological advancements is then described, followed by an assessment of current funding support for research and development on desalination.

STATUS OF DESALINATION USE

Separation of salt from water has a long history, dating from the time when salt, not water, was the precious commodity. As populations grew and demands for fresh water expanded, technologies were developed to produce fresh water in remote locations and on naval ships at sea. Sir Richard Hawkins reported in 1662 that during his voyages to the South Seas, he had been able to supply his men with fresh water by means of shipboard distillation (Birkett, 2003). In 1852, a British patent was issued for a distillation device (Simon, 1998). The island of Curaçao in the Netherlands Antilles was the first place to make a major commitment to desalination, and plants have operated there since 1928. In 1938, a major seawater desalination plant was built in what is now Saudi Arabia (Cooley et al., 2006).

Global desalination water production capacity has been increasing exponentially since 1960 to its current value of 42 million m^3/day, as seen in Figure 2-1. Of this global cumulative desalination capacity, approximately 37 million m^3/day is considered to be operational. This capacity includes seawater and brackish water desalination plants for municipal, industrial, agricultural, power, military, and demonstration ap-

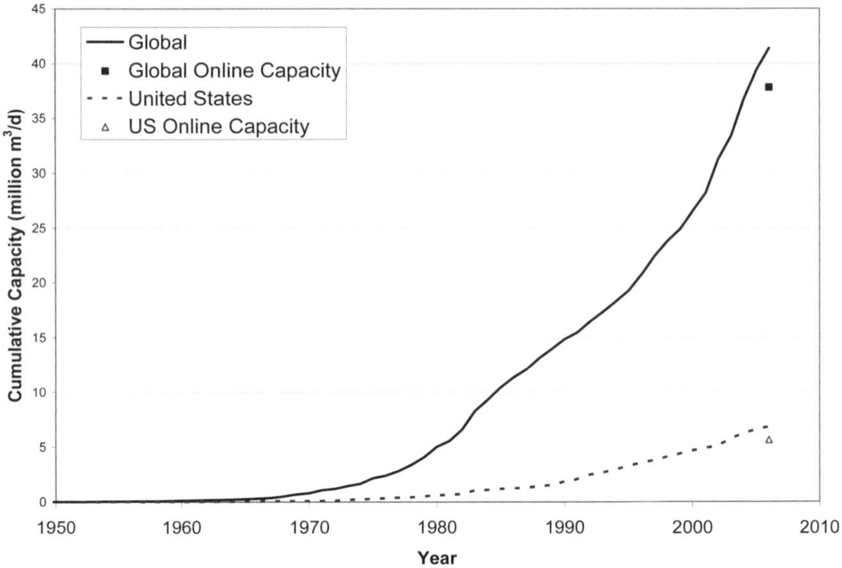

FIGURE 2-1. Cumulative capacity of installed desalination plants in the United States and worldwide from 1950 to 2006. The capacity of desalination plants that are online or presumed online in 2006 is shown as point data. Because this chart includes plants that have been decommissioned, the final cumulative capacity exceeds current operating capacity. Figure based on data taken from the 19th IDA Worldwide Desalting Plant Inventory (GWI, 2006b) and reproduced with kind permission of Global Water Intelligence.

plications, among others. These data were collected for the International Desalination Association's *Worldwide Desalting Plant Inventory*, which also includes facilities that use desalination technologies (e.g., reverse osmosis, nanofiltration) to remove salinity in the treatment of wastewater for reuse/reclamation, although reuse is not a focus of this report. The worldwide desalination capacity has approximately doubled since 1995 and continues to grow steadily. Nearly half (47 percent) of the current online global desalination capacity is located in the Middle East (Figure 2-2). North America, Europe, and Asia each have about 15 percent of the global online desalination capacity (GWI, 2006b).

The choice of desalination technology is a site-specific combination of many factors, including energy availability and form, source water quality, and other local conditions. Globally, thermal and membrane processes are the two major processes in use. In the United States, reverse osmosis and other membrane systems account for nearly 96 percent of U.S. online desalination capacity (see Figure 2-3) and 100 percent of the municipal desalination capacity (Mickley, 2006). Desalination tech-

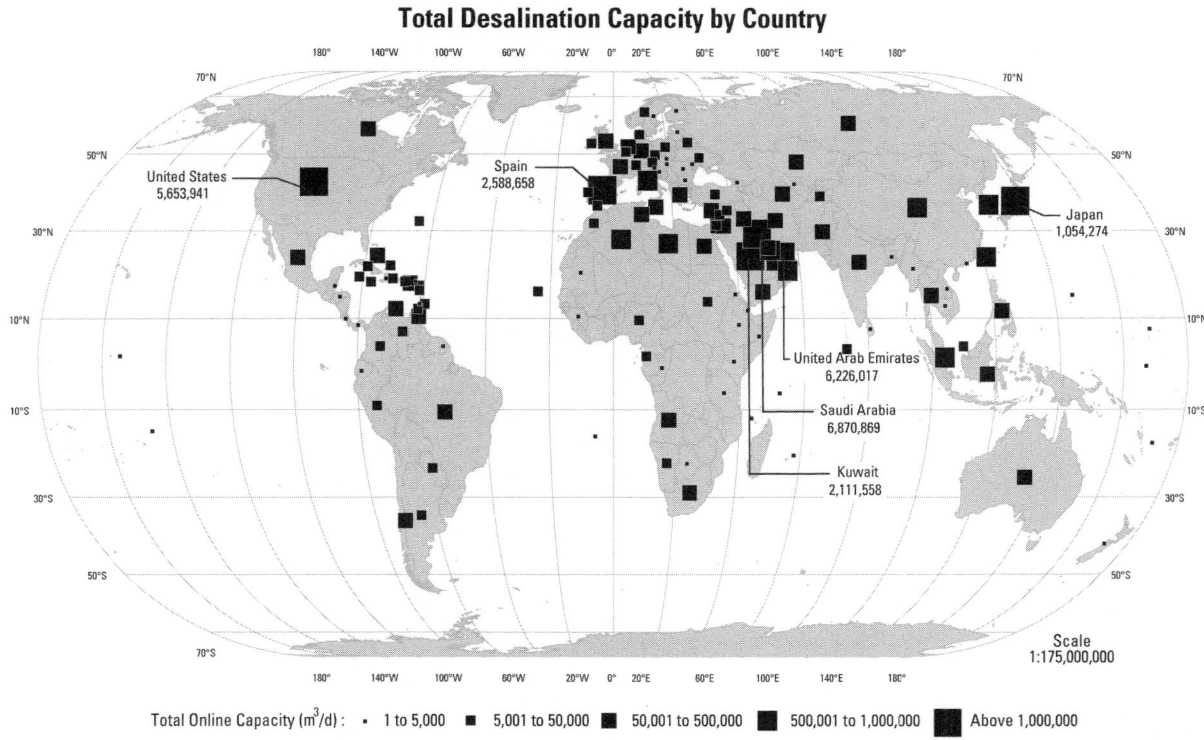

© International Mapping Associates

FIGURE 2-2. Global online desalination capacity.
SOURCE: Figure based on data from the 19th IDA Worldwide Desalting Plant Inventory (GWI, 2006b) and reproduced with kind permission of Global Water Intelligence.

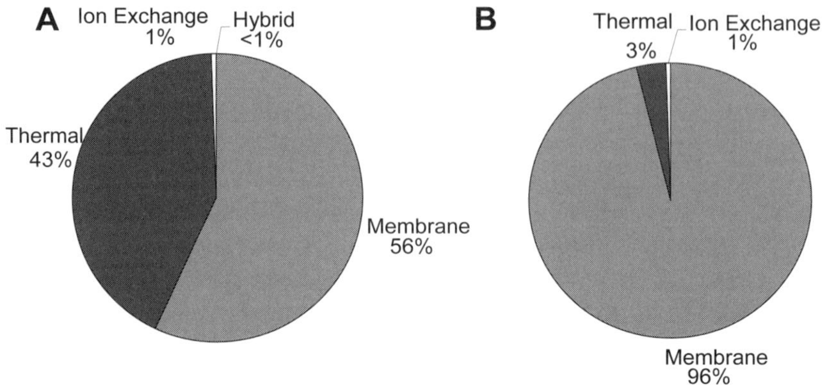

FIGURE 2-3. Percentage of total capacity of currently operating desalination plants by technology (a) worldwide and (b) in the United States (GWI, 2006b). These data include municipal and nonmunicipal (e.g., industry, power) desalination facilities. Figure based on data from the 19th IDA Worldwide Desalting Plant Inventory (GWI, 2006b) and reproduced with kind permission of Global Water Intelligence.

nologies are described in detail in Chapter 4.

Desalination plants have been built in every state in the United States, although nearly half of the plants are small facilities built for specific industrial needs. By 2005, approximately 1,100 desalination plants larger than 100 m^3/day (0.3 million gallons per day [MGD]) were online (or were presumed online). These plants have a total capacity of around 5.7 million m^3/day (1,500 MGD)—less than 0.01 percent of U.S. municipal and industrial water use (see Chapter 3). Between 2000 and 2005, the reported online desalination capacity in the United States increased by around 41 percent (Figure 2-1; GWI, 2006b). Three of the four states with the greatest installed capacity—Florida, California, and Texas—are coastal (see Figure 2-4). The fourth, Arizona, is an arid state with limited water supply sources. A large plant built by the U.S. government in Yuma, Arizona, to desalinate Colorado River water discharge is included in this estimate, but this plant has never operated outside of short test periods. Two-thirds of the U.S. desalination capacity is used for municipal water supply at roughly 300 facilities (Figure 2-5). Industry is also a sizeable user of desalination in the United States, with 18 percent of the national desalination capacity.

Seawater desalination reflects only a small portion (8 percent) of the

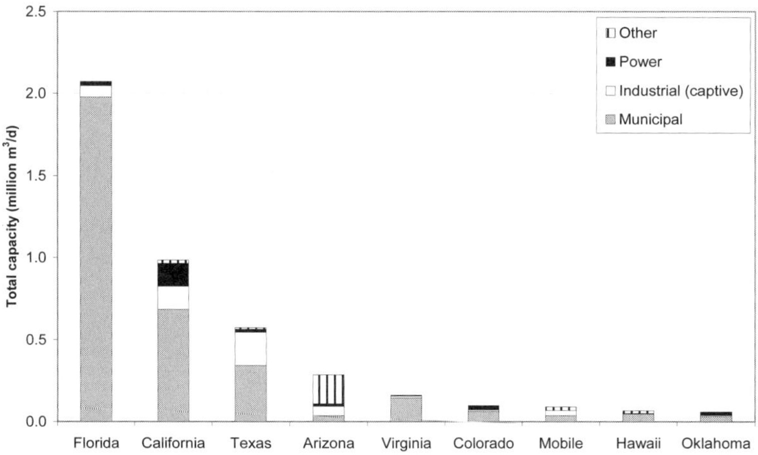

FIGURE 2-4. States with more than 1 percent of the U.S. current desalination capacity considering plants that are currently operating or are presumed to be operating. The total reported capacity is subdivided by end user according to four categories: municipal, industrial, power, and other (including military, irrigation, discharge, tourism, and demonstration).
SOURCE: Figure based on data from the 19th IDA Worldwide Desalting Plant Inventory (GWI, 2006b) and reproduced with kind permission of Global Water Intelligence.

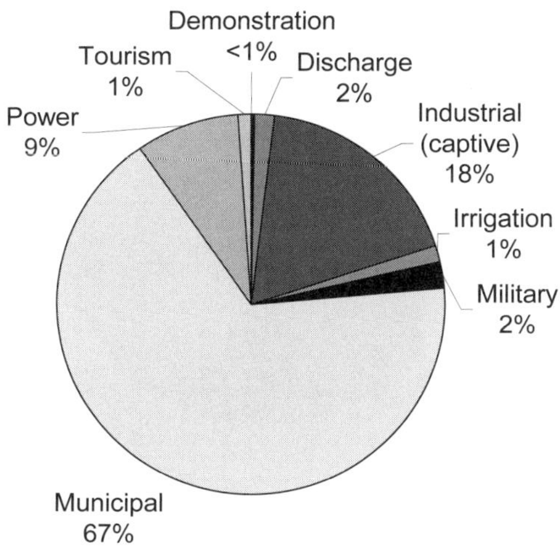

FIGURE 2-5. Percentage of total capacity of currently operating U.S. desalination plants by end user.
SOURCE: Figure based on data from the 19th IDA Worldwide Desalting Plant Inventory (GWI, 2006b) and reproduced with kind permission of Global Water Intelligence.

online capacity in the United States, but globally, 60 percent of the online desalination capacity relies upon seawater as source water (GWI, 2006b). Instead, the United States currently uses desalination technologies primarily to treat brackish water (77 percent of online capacity; see Figure 2-6). The remaining capacity is primarily dedicated to desalinating wastewater and providing pure water for high-quality industrial purposes.

Desalination processes can create potable water from many different source water qualities. The water quality of seawater varies from location to location, but average seawater contains mostly chloride and sodium ions and also low concentrations of ions such as bromide and boron, which can be potentially troublesome in membrane desalination processes (Table 2-1; see also Chapter 5). Brackish waters vary greatly in ionic composition across the country depending on their hydrogeologic origin. Table 2-1 reveals some of the variation in three brackish groundwaters and one surface water, but they are not necessarily representative of the possible variations. Each site listed in Table 2-1 is used as the source water or is near the source for an active desalination plant or one that is under consideration. Total dissolved solids concentration varies in these waters, as does the concentration of minor ions such as selenium,

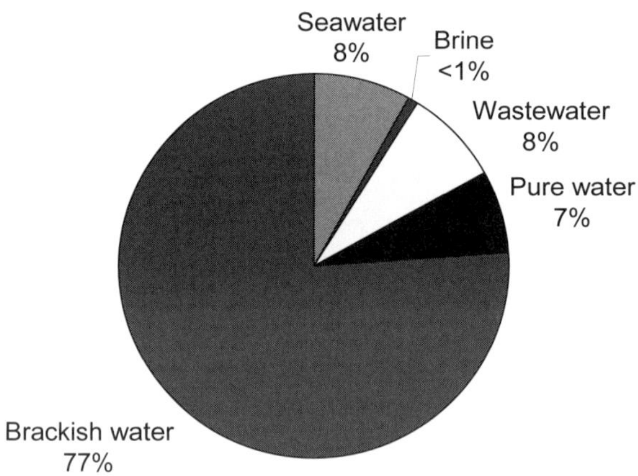

FIGURE 2-6. Percentage of total capacity of currently operating U.S. desalination plants by source water (GWI, 2006b). In this data set, brackish water is defined as water with total dissolved solids (TDS) between 500 and 15,000 mg/L, and pure water is that with TDS below 500 mg/L. Figure based on data from the 19th IDA Worldwide Desalting Plant Inventory (GWI, 2006b) and reproduced with kind permission of Global Water Intelligence.

TABLE 2-1. Examples of Variation in Water Quality Used as Source Water for Desalination

Ion (mg/L)	Average Seawater[a]	El Paso Water Utilities, TX Airport Wells[b]	Indian Wells Valley Water District, CA*[c]	Sarasota County, FL Well[d]	Colorado River Water near Andrade, CO[e]
TDS	35,000	3,170	1,630	1,180	1,021
Chloride	19,000	1,370	236	27.1	181
Sodium	10,500	745	333	24.6	185
Sulfate	2,700	301	570	609	342
Magnesium	1,350	38.4	49	70.1	38.3
Calcium	410	176	164	166	104
Potassium	390	15.9	6.1	4.02	5.7
Bicarbonate	142	75	370	144	160
Bromide	67	0.05	-	-	-
Strontium	8	-	1.55	-	1.4
Silica	6.4	29.4	45	-	14.2
Boron	4.5	-	1.74	-	-
Fluoride	1.3	0.61	1	-	0.5
Nitrate	3.0	0.11	72	-	2.6
Arsenic	0.003	-	0.0052	-	0.0035
Uranium	0.003	-	0.080	-	0.0038
Selenim	0.00009	-	0.059	-	0.0023

* Equal blend of four wells.
- No data.
SOURCES: [a] Hem, 1986; [b] J. Balliew, El Paso Water Utilities, personal communication, 2007; [c] Yallaly et al., 2007; [d] Brustlin, 2007; [e] USGS, 2006.

arsenic, and silica, posing unique challenges to the design of desalination processes. In some cases the challenge is in creating a quality product water; in others the challenge is safely and cost-effectively disposing of the resulting concentrate.

HISTORY OF DESALINATION RESEARCH AND DEVELOPMENT

The notable increase in the use of desalination over the past 40 years (Figure 2-1) is to a great extent the result of a long history of research and development efforts. Early research on desalination was conducted during World War II to satisfy freshwater needs in remote locations, and the United States and other countries continued that work after the war (Cooley et al., 2006). A major effort funded by the U.S. government began with the Saline Water Conversion Act of 1952, which established the Office of Saline Water (OSW). Housed in the Department of the Interior, the OSW funded research aimed at developing processes to recover drinking water from the oceans and brackish groundwater sources. Basic and applied research projects were initiated by the OSW with universities, institutions, and private companies on a wide array of subjects. The projects investigated basic fundamentals, the viability of recently developed processes, and new concepts. In 1974, the OSW became the Office of Water Research and Technology (OWRT).

Many of the early projects centered on thermal processes, both evaporative and freezing. Significant work was completed on materials of construction, heat transfer surfaces, corrosion, de-misters, and others. The work was instrumental in assisting the design and construction of some of the first large evaporative desalination systems in the Middle East. The OSW also invested funds into innovative desalination research, such as the development of the Zarchin process (freeze desalination; see Box 4-6) (J. Birkett, Westneck Strategies, personal communication, 2006).

During the late 1950s, the OSW funded basic work on the cellulose acetate polymer, the first practical membrane to be developed for desalination. In 1960, Sourirajan and Loeb made the first practical discovery of pressure-driven membrane technology to desalt water (Loeb and Sourirajan, 1963). OSW-funded research also led to the development of the solution-diffusion model of membrane transport, providing a theoretical basis for further advances (Lonsdale et al., 1965). The federal government sponsored a substantial amount of research by companies in development of heterogeneous cellulose acetate, cellulose triacetate, and hollow fiber membranes (both asymmetric and thin film). Federal research and development investments of the 1960s also spawned the commercial use of thin-film composite reverse osmosis technology through the 1970s. In the late 1970s, Riley and Cadotte embarked simultaneously on thin-film composite membranes that would later provide a giant step forward in increased membrane permeability at lower pressure for reverse osmosis (Riley et al., 1976; Cadotte, 1977).

During these halcyon years, the United States was looked upon as the undisputed leader in desalination technology. The OSW and the OWRT together spent more than \$1.5 billion in 2006 dollars and produced more than 1,200 technical reports (see Figure 2-7).[1] This government funding was responsible for the greatest development period in the growth of desalination technology. However, the OSW and the OWRT funded research that was focused on more than just desalination, includ ing general aspects of saline water research, such as the physical properties of saline fluids, hydrocarbon hydrates, and organic solutions. Such

[1] For more information, see http://www.usbr.gov/pmts/water/desalnet.html.

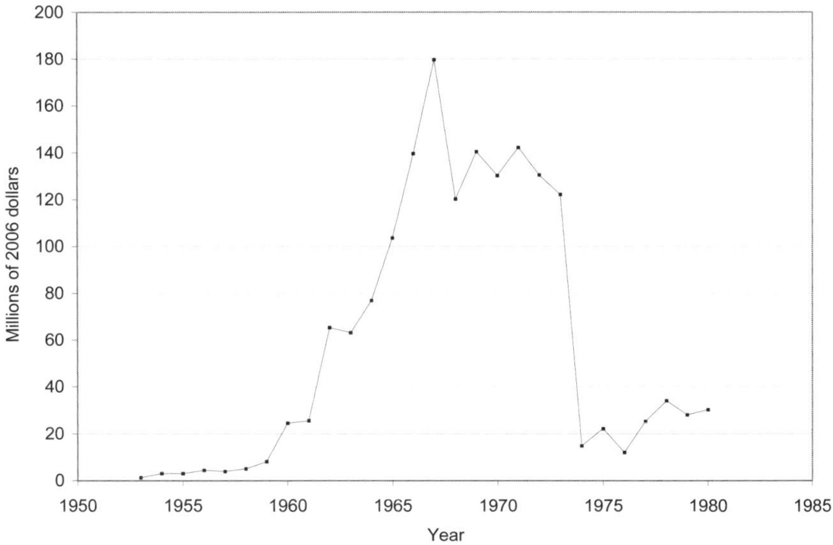

FIGURE 2-7. Yearly federal funding for desalination research and development between 1953 and 1980, as appropriated, in constant 2006 dollars. Based on data from the U.S. General Accounting Office (1979) and the Bureau of Labor Statistics Consumer Price Index.

reports are applicable to traditional water treatment processes and also led to the development of nonwater applications. Examples of non-water-related developments that came directly from or were enabled by OSW/OWRT technology include kidney dialysis and flue gas desulfurization technology to reduce air pollution. Entrepreneurial companies using OSW/OWRT technologies and reports supported the dramatic expansion of the reverse osmosis membrane market starting in the 1980s.

In the mid-1970s, federal research funding began to decline, and in 1982 federal funding for desalination research and development was discontinued except for a small amount of research in the Department of the Interior. The Water Resources Research Act of 1984 continued some desalination research in the U.S. Geological Survey, but the OWRT was closed. Private industry continued research and development with its own funds, although only the largest companies could engage in major research.

Twelve years later, mainly due to the efforts of Senator Paul Simon, Congress passed the Water Desalination Act of 1996 (P.L. 104-298) to renew federal research and development in desalination through grants, cooperative agreements, and in-house research. The purpose of the program was to determine the most technologically efficient and cost-effective means by which useable water could be produced from saline or contaminated water. The Act authorized program funding of $5 million per year for research and studies for 6 years, beginning with fiscal year (FY) 1997, for the Desalination and Water Purification Research and Development Program (DWPR) in the U.S. Bureau of Reclamation (USBR). In addition, $25 million was authorized over 6 years for demonstration and development (Mielke, 1999). The authorized funding was not fully appropriated, and the Act has been extended until 2011. Approximately $13 million was expended by the USBR on desalination research through the DWPR program between 1998 and 2006 (C. Hennig, USBR, personal communication, 2007). The USBR also supports desalination research and development through its science and technology program, its water reclamation and reuse (Title XVI) program, and its Water Quality Improvement Center in Yuma, Arizona, which provides facilities and assistance for pilot plant studies. Other governmental funds have been provided by congressional earmarks (or write-ins), such as the efforts to create the Brackish Groundwater National Desalination Research Facility in Alamogordo, New Mexico, which received over $20 million in funding through 2006 (C. Hennig, USBR, personal communication, 2007).

Separate efforts have been funded by the Department of Energy, the U.S. Army, the Office of Naval Research (ONR), and other governmental organizations, each concentrated on specialty areas of desalination to meet their organizational needs. Current research investments are described in more detail in the next section.

CURRENT DESALINATION RESEARCH FUNDING AND OVERSIGHT

Future direction for advancing and implementing desalination requires an understanding of the current level of research funding and oversight. To this end, a survey of U.S. federal agencies, state agencies, and nonprofit organizations was performed for FY 2005 to 2007 (see Appendix A for a copy of the survey document). Survey participants were asked to provide data on the level of expended funds for their agency's or organization's desalination research and to list separately any funding they provided for desalination construction and for water reuse research. Federal agencies were also queried on the percentage of that funding provided in the form of a congressional write-in as opposed to that which was requested in the president's budget for that year. All entities were asked to provide the percentages of desalination funding in basic research, applied research, and development according to the definitions of the Office of Management and Budget (OMB) (see Box 2-1). All entities were also queried on the percentages of their research activities that they considered to be high or low risk (as defined by the agency themselves) and short (< 3 years) or long term (> 3 years).

Federal funding of desalination research as reported in the survey is summarized in Table 2-2. The single largest federal sponsor of desalination research funding in recent years has been the USBR. The ONR, Sandia National Laboratory, and the National Science Foundation (NSF) have also expended significant funds in support of desalination research. The majority of federally funded desalination research is considered to be either applied research or development. Only NSF reports 80 percent of their research to be basic research, although ONR and several of the national laboratories also support some basic research. A significant percentage of the research funded through federal channels is considered by the agencies to be high-risk and long-term research. Examples of federal research currently under way include the following:

- developing a forward osmosis water purification prototype, by the USBR;
- investigating computational fluid dynamics for advanced membrane design, by Sandia National Laboratory;
- demonstrating novel components for use in desalination, through the ONR;
- quantifying the fresh and saline groundwater resources of the Salt Basin in New Mexico, by the U.S. Geological Survey;

BOX 2-1
Research Definitions

The Office of Management and Budget defines three categories of research (OMB, 2003):

Basic Research—"Basic research is defined as systematic study directed toward fuller knowledge or understanding of the fundamental aspects of phenomena and of observable facts without specific applications towards processes or products in mind."

Applied Research—"Applied research is defined as systematic study to gain knowledge or understanding necessary to determine the means by which a recognized and specific need may be met."

Development—"Development is defined as systematic application of knowledge or understanding, directed toward the production of useful materials, devices, and systems or methods, including design, development, and improvement of prototypes and new processes to meet specific requirements."

TABLE 2-2. Significant Federal Desalination Research Funding in Fiscal Years 2005-2007

	Research and Development Funding Expended (FY 2005 and 2006) or Appropriated (FY 2007) (millions)			Percent of Funding as Earmarks			Research Type (percent)				
Agency	FY 2005	FY 2006	FY 2007	FY 2005	FY 2006	FY 2007	Basic	Applied	Develop.	Percent High Risk[a]	Percent Long Term
Army	0.8	0.8	1.0	0	0	0	0	0	100	0	0
Argonne Natl. Lab	0.07	0.3	0.2	0	0	0	0	70	30	10	20
Lawrence Livermore Natl. Lab	0.1	0.1	0.1	0	0	0	100	0	0	70	100
Oak Ridge Natl. Lab	0	0.08	0.08	0	0	0	50	50	0	100	0
Sandia Natl. Lab	3.0	4.0	0	100	100	n/a	40	30	30	50	70
National Science Foundation	3.0	3.0	3.0	0	0	0	80	20	0	50	80
Office of Naval Research[b]	4.0	4.2	1.0	100	100	100	40	40	20	40	80
Bureau of Reclamation	11.8	10.9	3.9	60	57	0	0	60	40	60	60
U.S. Geological Survey	0.8	0.9	0.9	0	0	0	0	84	16	16	100
Total	23.6	24.2	10.1								

NOTE: The Environmental Protection Agency, Los Alamos National Laboratory, Tennessee Valley Authority, National Institute of Environmental Health Sciences, and National Energy Technology Laboratory were also surveyed, but they reported that no funds were expended or appropriated for desalination research for these agencies in FY 2005-2007; n/a = not applicable.

[a] Each agency determined its own definition of high risk research.

[b] The ONR also has additional funding ($7.7M, 4.5M, and 2.3M for FY 2005, 2006, and 2007, respectively) for development of the Expeditionary Unit Water Purification, a 379 m^3/day (0.1 MGD) portable water desalination and water treatment unit for military deployment to remote field locations or to civilians in cases of water emergencies. These demonstration funds are not included here in the research and development funding totals.

- developing processes to remove marketable mineral by-products from concentrate for inland desalination, through Lawrence Livermore National Laboratory;
- developing new chemicals to more effectively clean desalination membranes, through Argonne National Laboratory; and
- developing active fouling-resistant nanofiltration and reverse osmosis membranes, through the NSF.

Federal appropriations in desalination research declined significantly in FY 2007 as compared to FY 2005 and FY 2006 (see Table 2-2). This is at least partly attributable to the absence of congressional earmarks associated with continuing resolutions under which most agencies were operating in FY 2007. The absence of earmarks most significantly affected the budgets of the USBR, ONR, and Sandia National Laboratory. The importance of earmarks in determining the magnitude of federal spending on desalination research underscores the fact that such federal investments are not integrated and that the resulting research and development are not conducted in a strategically purposeful fashion. Thus, there is a strong need for a strategic and coordinated approach to federal desalination research.

Several states, most notably California, have begun to direct their own budgetary resources toward desalination research and development to meet their immediate water supply concerns. Florida and Texas also sponsor such research, although on a much smaller scale, as shown in Table 2-3. California's efforts, pursuant to Proposition 50, a publicly approved referendum, provided nearly $50 million for desalination research, development, and construction in 2005-2006 (state construction funding is listed separately in Table 2-5). This program is one of the largest publicly funded desalination research programs ever established. The states report that the research that they fund is largely low risk and focused on the attainment of short-term results. Research topics currently supported through state funding include the following:

- Location-specific feasibility evaluations for desalination concentrate management,
- Project management and technical support services for desalination and concentrate management studies,
- Location-specific feasibility of co-locating seawater treatment facilities with power plants, and
- Concentrate reduction/zero liquid discharge demonstration.

TABLE 2-3. Significant State Desalination Research Funding in Fiscal Years 2005-2007

	Research and Development Funding Expended (FY 2005 and 2006) or Appropriated (FY 2007) (millions)			Percent Research by Type			%	%
State	FY 2005	FY 2006	FY 2007	Basic	Applied	Develop.	High Risk	Long Term
CA	14.0	11.9	0.4	0	17	83	7	0
FL	0.4	0.5	0.4	0	40	60	0	25
TX	0.8	1.6	1.6	0	12	88	0	0
Total	15.2	13.9	2.4					

NOTE: The Florida agencies included in this table are the South Florida Water Management District and the St. Johns River Water Management District. The California agencies/programs included in the above results are the Proposition 50 program, the California Energy Commission, and the State Water Resources Control Board of California, which supports desalination research through the WateReuse Foundation. Construction funding provided through the Proposition 50 program is tallied separately in Table 2-5. Texas funding comes through the Texas Water Development Board.

Nonprofit foundations and institutes such as American Water Works Association Research Foundation (AWWARF), the Water Environment Research Federation (WERF), and the National Water Research Institute (NWRI) have also provided funding, albeit at modest levels as shown in Table 2-4. The WateReuse Foundation also sponsors desalination research, but those funding data are not tabulated here because the funding comes from other agencies and is already accounted for elsewhere in this survey. The data suggest that few nongovernmental organizations (NGOs) are involved in desalination research and that the research funding they provide is fairly small. Less than $0.8 million is expended annually by these foundations combined. Foundation-sponsored research is generally short-term, low-risk, applied research, although NWRI supports some basic and high-risk research. Examples of research supported through these nonprofit organizations include the following:

- Investigation of regional solutions for disposing of concentrate,
- Zero liquid discharge and volume minimization for inland desalination,
- Feed water intake systems for desalination plants,
- Desalination facility design and operation for maximum energy efficiency,
- Development of smart nanofiltration membranes, and

- Crystallization to enhance two-stage reverse osmosis recovery.

Further comparisons are possible from the survey results. Nearly all funding for basic research (as defined by OMB) comes from the federal government, although the federal government also provides significant funding toward applied research and development. Nonprofit organizations tend to support applied research, and state funds primarily support development projects, including feasibility studies and pilot testing.

Table 2-5 provides a 3-year comparison for FY 2005-2007 of total federal, state, and nonprofit foundation funding for desalination research, desalination construction, and funding for water reuse research. The significant drop in desalination research funding provided by federal sources is accompanied by a similar decrease in desalination construction funding, reflecting the impact of the continuing budget resolutions on desalination research funding. State funding appears to be moving away from research and into construction in this 3-year scenario, although this scenario was strongly affected by California's Proposition 50 funds in 2005 and 2006. In 2007, state construction appropriations exceeded research appropriations by a factor of more than 15. Water reuse research funding has been holding fairly stable across all funding providers at around $15 million annually. This amount is less than desalination

TABLE 2-4. Significant Nonprofit Foundation Desalination Research Funding in Fiscal Years 2005-2007

	Research and Development Funding Expended (FY 2005 and 2006) or Appropriated (FY 2007) (millions)			Percent Research by Type			% High Risk	% Long Term
Foundation	FY 2005	FY 2006	FY 2007	Basic	Applied	Develop.		
AWWARF	0.51	0.06	0.25	0	100	0	0	15
NWRI	0.27	0.19	0.22	60	40	0	50	<10
WERF	0.02	0.01	0.01	0	100	0	0	0
Total	0.78	0.25	0.47					

NOTE: The Water Environment Research Federation (WERF) funding reflects that provided to the WateReuse Foundation in support of desalination research in FY 2005-2007. The WateReuse Foundation is not specifically included in this table, because their desalination research funding is primarily derived from other agencies that are already tallied elsewhere in this survey.

TABLE 2-5. Comparison of Funding from Federal, State, and Foundation Sources to Desalination Research, Desalination Construction, and Water Reuse Research in Fiscal Years 2005-2007

	Desalination Research Funding Expended (FY 2005 and 2006) or Appropriated (FY 2007) (millions)			Desalination Construction Funding Expended (FY 2005 and 2006) or Appropriated (FY 2007) (millions)			Water Reuse Research Funding (millions)		
	FY 2005	FY 2006	FY 2007	FY 2005	FY 2006	FY 2007	FY 2005	FY 2006	FY 2007
Federal	23.6	24.2	10.1	22.8	17.6	9.3	12.6	12.4	10.2
State	15.1	13.8	2.3	10.8	27.2	40.2	0.2	0.4	0.8
Foundation	0.8	0.2	0.5	0	0	0	1.6[a]	2.0[a]	4.5
Total	39.5	37.2	12.9	33.6	44.8	49.5	14.4	14.8	15.5

[a] Foundation funding totals in 2005 and 2006 were adjusted for earmarked funds included in federal totals.

research funding for FY 2005 and FY 2006, but slightly greater than desalination research funding in FY 2007.

Industrial Research and Development Funding

With the demise of OWRT (formerly OSW) in the early 1980s, research and development in desalination became primarily the responsibility of the private sector. Research and development expenditures for private industry tend to be unreported and are considered proprietary information, thus precluding a rigorous, documentable tally of industry research spending on desalination. Nevertheless, the committee used a simple and logical analysis, informed by the considered judgment of those working in the industry, to estimate this research spending. The committee estimates that private industry roughly invests between $100 and $150 million per year in research and development on desalination technology and its applications, far exceeding federal government research spending in desalination. This estimate was made by analyzing the annual reports of Siemans, GE (and the water treatment companies that they acquired), Veolia, Suez Degremont, Dow (Filmtec), Nitto Denko (Hydranautics), Toray, Pall, ITT, and Hyflux—companies that are estimated to represent over 75 percent of the desalination membrane supply market worldwide. For the purpose of this analysis, unless more detailed data were available, the committee assumed that the percent of revenue spent on research and development was equally allocated over all of a company's business segments.

Based on the judgment of individuals working in the industry, it is reasonable to assume that the majority of research and development funding in industry is directed at low- to moderate-risk projects and that as little as 10 percent of all research and development funding is allocated to high-risk research on novel products and processes, where the outcomes of the research are more uncertain. Based on this assumption, the committee estimates that industry spends roughly $10 to $15 million per year on high-risk research and development to develop proprietary desalination products. This amount of funding is of the same order of magnitude as the annual federal funding over the past 3 years. It should be noted that this estimate does not include venture capital funds, which are typically targeted at developing new technologies and bringing them to market once they are proven in the early research stage.

International Governmental Research and Development

The committee did not formally assess the magnitude of international funding for desalination research and development, but other countries like Singapore, China, Israel, the European Union, and the United Arab Emirates also fund desalination research and development. For instance, ongoing research programs exist at the Kuwait Institute for Scientific Research Doha Research Plant (KISR, 2007), the Research and Development Center of the Saline Water Conversion Corporation in Saudi Arabia (Al-Sofi, 2001), and the Middle Eastern Desalination Research Center (MEDRC, 2007). The European Commission recently provided nearly 4.7 million Euros for research and development efforts in membrane desalination and solar-powered desalination (European Commission, 2007). The majority of these research and development efforts are focused on improving the desalination processes that use today's best available technologies. The most significant government-funded research and development investment is in Singapore, where a dedicated office, the Environment and Water Industry Development Council (EWI), has been set up to spearhead the growth of the environment and water industry. The launch of the EWI is to be supported by Singapore's Research, Innovation and Enterprise Council, which will provide $330 million of research and development funds over the next 5 years (2006-2011) to catalyze the development of the local water industry by funding applications development and high-risk research and development.[2] It is not known exactly what portion of these funds will be focused on desalination, but these funds are anticipated to be comparable to or greater than current U.S. government desalination research funding.

---

[2] http://app.mewr.gov.sg/press.asp?id=CDS4096.

CONCLUSIONS AND RECOMMENDATIONS

The history of desalination is centuries long. Following early research and development efforts, there has been an exponential increase in desalination capacity installed both globally and nationally since 1960. Desalination plants, for purposes ranging from municipal water supply to industrial applications, are now in place in every state in the United States. These plants primarily utilize membrane technology and treat mainly brackish water rather than seawater.

The exponential growth in desalination in the United States has been made possible by federal funding of more than $1.5 billion in today's dollars for desalination research and development from the late 1950s to the early 1980s. More recently, the Desalination and Water Purification Research and Development program in the USBR has been the major federal force in desalination research and development, supplemented by significant federal earmarks, although nonfederal funding is increasingly important.

The committee's survey of federal, state, and NGO investments in desalination research and development together with the estimates and analyses of private-sector funding lead to several conclusions.

There is no integrated and strategic direction to the federal desalination research and development efforts. Desalination research and development efforts are funded through at least nine federal agencies and laboratories, each with their own research objectives and priorities. The majority of federal desalination research and development funding also comes from congressional earmarks, which limits the ability to develop a steady research program. Federal funding for desalination research declined by nearly 60 percent from FYs 2005 and 2006 to FY 2007, largely due to an absence of earmarks in FY 2007. The results of the survey underscore the need for the development of a national strategic research agenda for desalination. Recommendations to address these concerns are outlined in Chapter 8.

Anecdotal evidence suggests that international investments in desalination research and development are greater than current research investments from the U.S. government. Desalination technology is being used to meet water supply demands in over 140 countries around the world, and international support for desalination research appears to be growing. Singapore recently announced a major research initiative to catalyze the development of the country's water industry.

State governments, especially that of California, have made sizeable recent investments in desalination research and development.

The majority of this funding is directed at site-specific or region-specific problems, with heavy emphasis on pilot and demonstration projects. State spending on desalination plant construction has grown rapidly in the past 3 years and has greatly overshadowed spending on desalination research.

The private sector appears to fund the majority of desalination research, with total annual spending estimated to be more than twice that of all other surveyed sources of such funding. Based on the judgment of individuals working in the industry, the private sector allocates a smaller fraction of its research portfolio to high-risk desalination research than the federal government. Given the large research and development budgets in private industry, however, high-risk research funding is estimated to be roughly equivalent between the private sector and the government.

3

Water for the Future

Decisions about investments in desalination technologies should be made in a comprehensive and consistent manner that adequately considers a wide range of economic, environmental, and political factors. Among the most important factors are how future demands for water are to be evaluated or forecasted, and how those projected demands are to be satisfied given a wide range of water supply alternatives. The past few years have seen a growing interest in, and worry about, the issue of water scarcity in arid and semi-arid regions of the United States. That concern has now spread to the more humid regions of the country, and the availability of water to satisfy growing demands for domestic, agricultural, and environmental uses is now a matter of increasing concern in virtually all regions. This chapter assesses the various forces and factors affecting supply and demand for water and emphasizes the ways in which those factors may affect the social attractiveness and economic competitiveness of desalination technologies.

Examples of the need to reevaluate water supply and demand abound. The City of Atlanta is struggling to find supplies to support its rapidly expanding urban areas, while Florida is concerned about using the same watersheds to maintain adequate flows to protect ecosystems and wetlands. The states of Maryland and Virginia needed the Supreme Court to help them resolve a dispute about use of the shared waters of the Potomac River (*Virginia v. Maryland,* 540 U.S. 129 [2003]). The allocation of the waters of the Missouri River basin between navigational purposes in support of agriculture on the northern plains and environmental uses, including support for a number of endangered species, is the subject of continuing and unresolved controversy (NRC, 2002b). Colorado River water management continues to be a focus of dispute among seven western states and Mexico (NRC, 2007). And perennial water problems in California have been heightened by a recent court decision that may reduce water pumping for agricultural and urban uses to protect the endangered delta smelt (Ricci and Bailey, 2007).

As population and regional economies grow in the future and as the importance of providing water to support environmental services becomes more widely appreciated, the overall pressure on the nation's limited water resources will continue to intensify. Simultaneously, interest in finding novel means of managing these pressures will also intensify as the limitations of traditional infrastructure solutions are becoming better understood.

In recent years, increased attention has been drawn to the promise and prospects of desalination technology for alleviating the growing water scarcity. At its simplest, the technology might substantially reduce water scarcity by making the almost inexhaustible stock of seawater and the large quantities of brackish groundwater that appear to be available into new sources of freshwater supply. Historically, the deployment and use of desalination technology has been constrained because of its high costs, and its use has been confined to places in the world where energy is cheap and alternative sources of supply are either unavailable or especially costly. However, recent advances in technology, especially improvements in membranes that can be used in desalinating both brackish water and seawater, have rekindled more widespread interest in desalination. Indeed, the total costs of desalinating brackish water and seawater (including concentrate management costs) in the United States may now be competitive with other alternatives in some locations and for some high-valued uses, fueling optimism over the prospects for the expanded use of desalination technology.

Seawater desalination technology is unique among supply augmentation alternatives in that it is not dependent on the hydrological cycle and can produce water as reliably during drought events as at other times. Brackish groundwater desalination can also provide reliable water supplies during short-term droughts, although longer periods of drought can affect regional groundwater availability. The recent occurrence of severe droughts in many areas of the country and the prediction that global climate change will likely result in regional increases in the frequency and intensity of extreme events have heightened public sensitivity to the variability of the hydrologic cycle. Consequently, desalination technology is attractive both because it offers the possibility of supplying large amounts of new water and because it can do so reliably. There is considerable interest, therefore, in advancing desalination technology and concentrate management alternatives and hastening the time when the costs are routinely competitive with the costs of other alternatives. At the same time, too heavy a reliance on any one source of water also imposes risks and vulnerabilities.

This chapter reviews historical and current thinking about the concept of water availability, use, scarcity, and sufficiency. Trends in U.S. and regional water use are presented along with a review of experience with water demand forecasts. The role of water planning and projections is also discussed in the context of assisting decisions about long-term investment in technologies such as desalination and nontechnological solutions such as demand management and water efficiency improvements.

ELEMENTS OF WATER SUFFICIENCY

During the nineteenth and twentieth centuries, the demand for water rose dramatically with growing populations and economies. Billions of dollars were spent on thousands of large water projects designed to control floods, move water vast distances, disinfect and treat water supplies, and deliver water for irrigation or hydropower. Even today, worldwide, tens of billions of dollars are spent annually on new dams (World Commission on Dams, 2000). Today, many of the largest cities of the United States rely on water brought from hundreds and even thousands of kilometers away. Agricultural productivity is highest where irrigation mitigates the effects of climatic variability, such as in California's Central Valley. The growing discussion about expanding investments in desalination technology reflects the perception that demands for water are growing in places where natural availability is constrained by environmental, financial, or social factors.

In recent years, however, a shift away from the construction of new water infrastructure and toward rethinking the active management of water use has started to occur, in part because of the improved understanding of the true financial, social, and environmental costs of large infrastructure, and in part as a result of new thinking about how best to meet the water requirements of human needs and desires (e.g., Douglas et al., 1998; Gleick, 2003a). New water facilities are still needed in many parts of the world, and the existing infrastructure must be maintained in order to keep the flow of benefits coming. But water planners and managers, as well as those responsible for using and paying for water, are now beginning to evaluate the other side of the equation—how society uses and manages water.

People require modest amounts of water for drinking, cooking, cleaning, and hygiene to maintain human well-being. For certain requirements, the actual "demand" for water may be unrelated to the minimum amount of water "needed" for that required benefit. Water demand to flush a toilet can range from 23 liters (6 gallons) in an old, inefficient U.S. toilet, to 5 liters (1.6 gallons) in a model that meets current U.S. standards, to zero liters in an efficient electric or composting toilet. New digital photography provides a faster, more detailed image while completely eliminating the water used by old film processing. In these examples, what is actually being demanded is not a specific amount of water but reliable services of removing wastes or producing goods and services (Gleick, 2003a). In addition, as income grows, people want recreation, leisure, and luxury goods. Providing these needs and wants can be accomplished in many ways that depend on technology, prices, cultural traditions, and other factors, often with radically different implications for water. Increasingly, water providers are exploring ways to deliver diverse water services matched to the users' needs and are working with water users at local and community scales. These changes in water policy, planning, and management have resulted in a dramatic change in the relationship between water withdrawals and population growth.

Review of U.S. Water Use over Time

Beginning in 1951, the U.S. Geological Survey (USGS) has published a series of comprehensive reports on water use in the United States at approximately 5-year intervals (e.g., Hutson et al., 2003; MacKichan, 1951, 1957; MacKichan and Kammerer, 1961; Solley et al., 1998). These assessments include estimates of surface and groundwater use separated into various water-use categories for all 50 states and by major hydrologic regions. (Definitions of "use" vary across studies and common vernacular; therefore, some standard definitions are provided in Box 3-1.) The initial study in this series estimated water use for all withdrawals, including municipal, rural, domestic, livestock, irrigation, industrial, and hydroelectric power. Water for in-stream flows such as navigation, recreation, and fish and wildlife were also addressed, though only qualitatively. Consumptive use of water began to be estimated in the 1960 report, and estimates were made of water use in the early decades of the century, beginning in 1900 (CEQ, 1991; MacKichan, 1957; NRC, 2002a). The most recent report, however, has eliminated several categories of information on water use. Consumptive use, in particular, is no longer reported, which will make tracking some of the most important impacts more difficult.

There are significant differences in the depth and quality of water-use data because of differences in types of water use, methods of data collection, reliability of reporting, and funding priorities (DOI, 2002). As a result, the National Research Council (NRC, 2002a) recommended a series of improvements in the U.S. National Water-Use Information Program to help provide more comprehensive and accurate water-use data in the future (see Box 3-2).

BOX 3-1
Definitions of Water Use, Conservation, and Efficiency Terms

There is confusion in the water literature about the terms use, need, withdrawal, demand, consumption, and consumptive use of water. The term *water use*, while common, can mean many different things, referring at times to consumptive use and at times to withdrawals of water. *Withdrawal* usually refers to water removed from a source and used to meet a human need. Some of this water may be returned to the original source with changes in the quantity and quality of the water, but some may be used consumptively. The term *consumptive use* or *consumption* typically refers to water withdrawn from a source and made unavailable for reuse in the same basin, such as through conversion to steam, losses to evaporation or transpiration, seepage to a saline sink, or contamination. Consumptive use is sometimes referred to as irretrievable or irrecoverable loss. Thermoelectric power plants typically withdraw substantial amounts of water for cooling but consume little, returning that water directly to the river, albeit warmer, for use by the next downstream user. A farmer may withdraw large amounts of water for irrigation, but the vast majority of it may be used consumptively by plants and be lost to any downstream users. *Need* for water is also a subjective term, but it usually refers to the minimum amount of water required to satisfy a particular purpose or requirement. The term *need* implies, usually incorrectly, that water is required to satisfy a purpose without regard to price. Thus, the term also sometimes refers to the desire for water on the part of a water user. *Demand* for water is an economic concept that is used to describe a want for water backed up by a willingness to pay. A demand schedule or curve (the graphical representation of demand as a function of price) summarizes the quantities and qualities of water that consumers are willing to take at different prices. Demand curves almost invariably show that the higher the price, the lower the quantity taken, though the slope of the curve varies with many factors (Boland et al., 1984).

Despite limitations in the data, these water-use studies have proven to be extremely valuable for both researchers and policy makers. One of the most important findings from the long-term data has been an unexpected change in the trend of U.S. water use. Both total U.S. freshwater withdrawals and gross national product exhibited rapid growth until the late 1970s (Figure 3-1). Although economic growth continued to rise exponentially, water withdrawals began to level off and even decline, a change not noted or recognized by water managers or policy makers until the 1990s. This decline, however, has persisted; indeed, it is even more apparent when per-capita use is measured (Figure 3-2). Since the late 1970s, per-capita withdrawals have declined nearly 25 percent and now are at levels comparable to those of the late 1950s and early 1960s. Yet U.S. population has grown from approximately 180 million in 1960 to over 300 million in 2006 (U.S. Census Bureau, 2007).

While no detailed analysis of the reason for these trends has been prepared, two major factors are relevant: changes in the efficiency of

BOX 3-2
Challenges to Reporting Water Use Data

Despite the value of consistent national water-use data and recommendations from the National Research Council for improvements in data collection (NRC, 2002a), recent cutbacks imperil the development of effective national water policy. The most recent USGS national water-use report (Hutson et al., 2003; based on data from 2000), for example, cut back on collection and presentation of data for several categories of use, including mining, livestock, and aquaculture. Data were only compiled for states where water uses in these categories are large. Withdrawals from major groundwater aquifers are only being reported for public supply, irrigation, and industry. Data on commercial water use, wastewater treatment, reservoir evaporation, and hydroelectric power are no longer being collected, nor is information on consumptive use, reclaimed wastewater, return flows, or deliveries from public suppliers. Researchers and water planners depend on many of these data to understand long-term trends in sectoral and regional water use

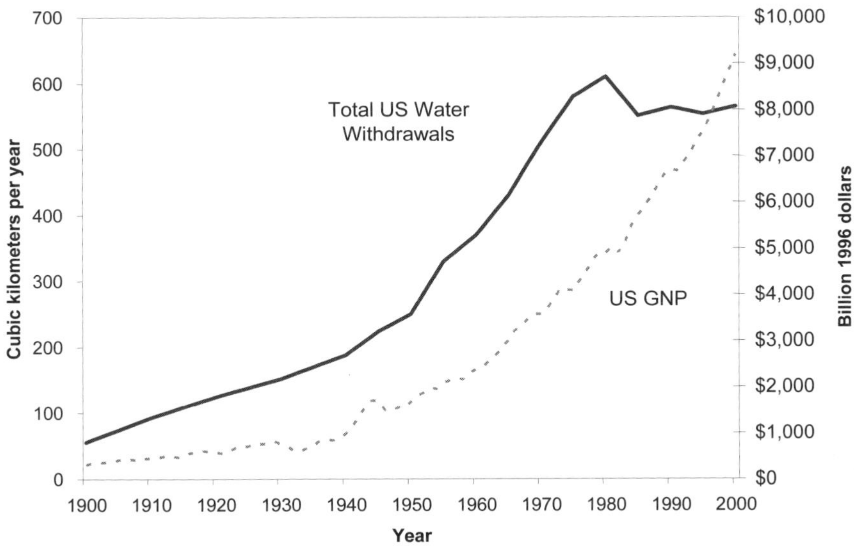

FIGURE 3-1. Gross national product (GNP, in billions of 1996 dollars) and total water withdrawals (cubic kilometers per year). Water withdrawal data from MacKichan (1951, 1957), MacKichan and Kammerer (1961), Solley et al. (1998), and Hutson et al. (2003). Economic data from Johnston and Williamson (2002). Data on water withdrawals before 1950 come from CEQ (1991).

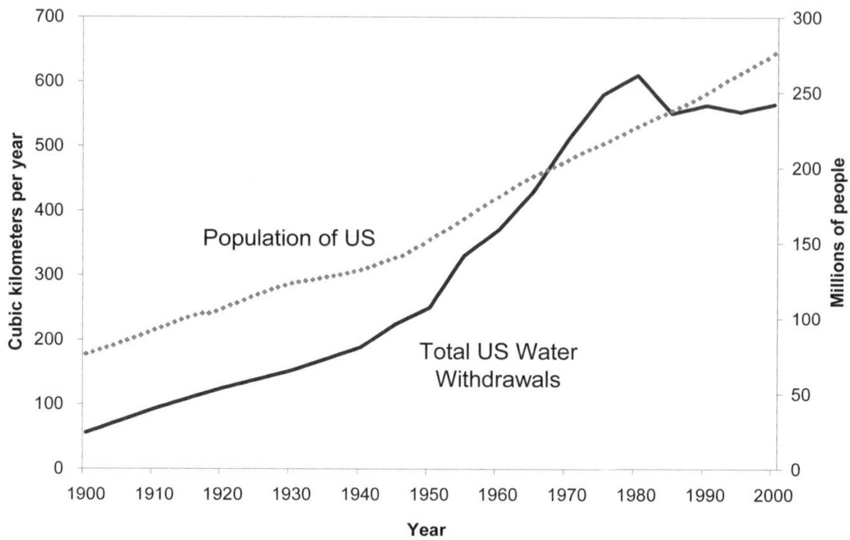

FIGURE 3-2. Total freshwater withdrawals in the United States from 1900 to 2000 (in cubic kilometers per year). Total water withdrawals based on data from MacKichan (1951, 1957), MacKichan and Kammerer (1961), CEQ (1991), Solley et al. (1998), and Hutson et al. (2003). Population based on data from U.S. Census Bureau (2001).

water use and changes in the structure of the U.S. economy (Gleick, 2003b). In the first case, significant improvements in technology have led to reductions in the amount of water required to produce specific goods and services. For example, new national appliance standards put in place in the mid-1990s have reduced the amount of water required to flush a toilet from an average of 6 gallons per flush to 1.6 gallons per flush. Similar standards have reduced flow rates of showerheads. Other factors have also played a role. Producing a ton of steel before World War II required 100 to 200 tons of water (Kollar and MacAuley, 1980; National Association of Manufacturers, 1950). By the turn of the millennium, competitive economic pressures and water quality discharge regulations had driven improvements in water-use efficiency in the best steel plants to less than 4 tons of water per ton of steel (Posco Steel, 2007). Changes in design and production processes also play a role: a ton of aluminum can be produced using only one and a half tons of water. Thus, as automobile production replaces steel with aluminum, as has been happening for many years, total water use per unit car drops further (Gleick, 2002).

Some differences in water-use trends can be seen in an analysis of water use in various sectors. Although overall water use in the United

States has remained roughly the same over the past 30 years, per-capita water use has decreased substantially (see Figures 3-1 and 3-2). Most of the magnitude of this decline comes from reductions in water withdrawals for irrigation, power plant cooling, and other industrial uses. There have been modest and regular increases in public water supply withdrawals in the United States, primarily for residential use, increasing approximately 2 million m^3/day per year (see Figure 3-3). Most reductions in agricultural water use have come about because of improvements in irrigation technology, including precision irrigation and better monitoring of soil moisture, both of which can reduce water use and increase the reliability and yield of crop production.

Changes in the nature of economies have played a role in reducing industrial water use (Figure 3-3). In the United States, the economy is less dependent on water-intensive industries such as mining, milling, and manufacturing and more a function of sectors that require less water per unit of output, such as telecommunications, computers, and services. Decreases in industrial water use are also the result of national legislation that set standards for the quality of wastewater discharges. One of the least expensive ways of meeting the federal water quality standards is to reduce the overall volume of water used, which led many industries to seek changes to the way process water was used.

Reductions in water withdrawals for thermoelectric power production have been driven by the gradual replacement of once-through cooling systems with systems that recycle water (Figure 3-3). This trend will

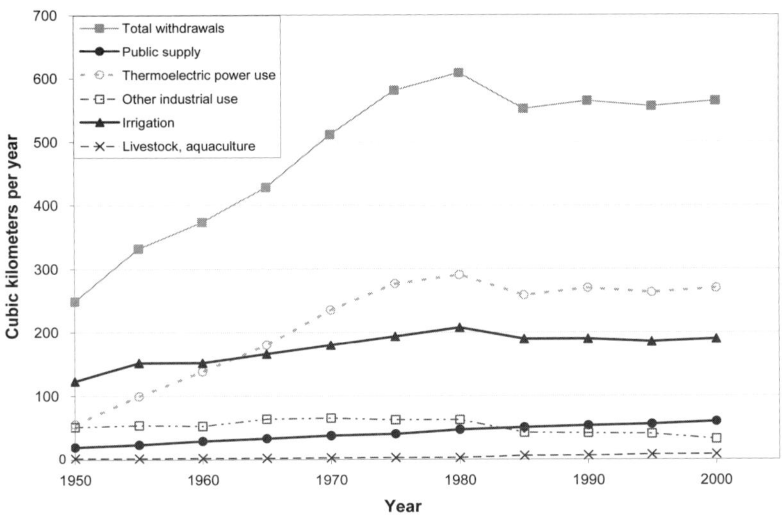

FIGURE 3-3. Changes in U.S. water withdrawals by sector. Data from MacKichan (1951, 1957), MacKichan and Kammerer (1961), Solley et al. (1998), and Hutson et al. (2003).

accelerate in the future as the Environmental Protection Agency requires more once-through cooling to be phased out (Clean Water Act Section 316b; see Box 5-1).

Details on changes in regional water use are also available from individual state data and from the USGS. As an example, Figure 3-4 shows long-term water withdrawals for California (1960 to 2000). In parallel with national trends, California has experienced substantial growth in population and economic indicators but declining per-capita water withdrawals (Hutson et al., 2003; MacKichan, 1951, 1957; MacKichan and Kammerer, 1961; Solley et al., 1998). Total water withdrawals in California have declined from a peak in 1980 and have remained roughly level since the mid-1980s, with an upward trend in water withdrawals in recent years. As with the rest of the United States, California irrigation withdrawals have dropped since 1980, and total urban water use has grown modestly since 1960 (Figure 3-4). Similarly, total water use in Texas has held relatively steady for over 25 years, although municipal water use increased steadily between 1974 and 2000 (Figure 3-5). In contrast, the City of Seattle has substantially reduced total water use since the late 1960s (see Figure 3-6) despite a 40 percent increase in population through comprehensive and consistent water efficiency programs (Seattle Public Utilities, 2007).

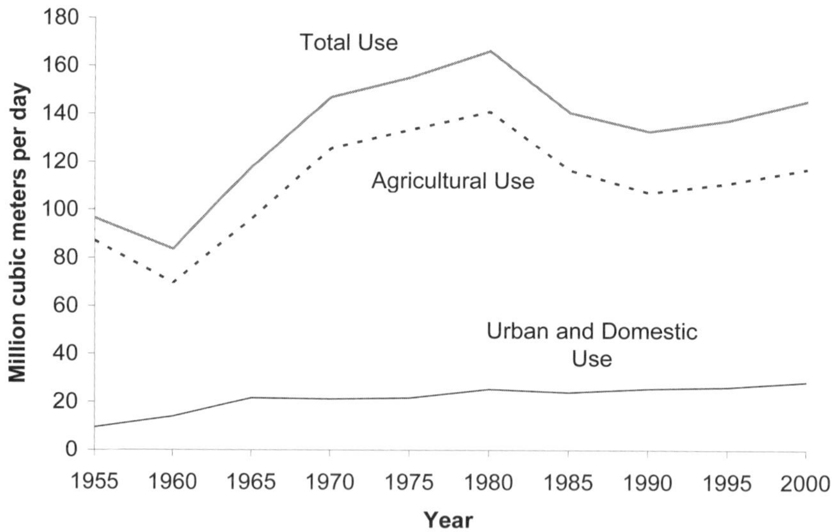

FIGURE 3-4. Water withdrawals in the state of California. SOURCES: Data from MacKichan (1951, 1957), MacKichan and Kammerer (1961), Solley et al. (1998), and Hutson et al. (2003).

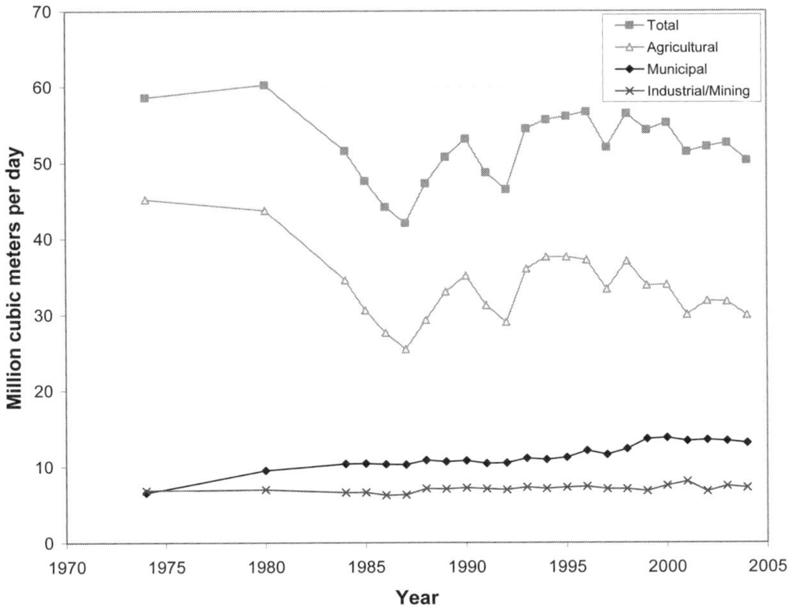

FIGURE 3-5. Texas water use, including both groundwater and surface water, between 1974 and 2004. SOURCE: Data from Texas Water Development Board (2007).

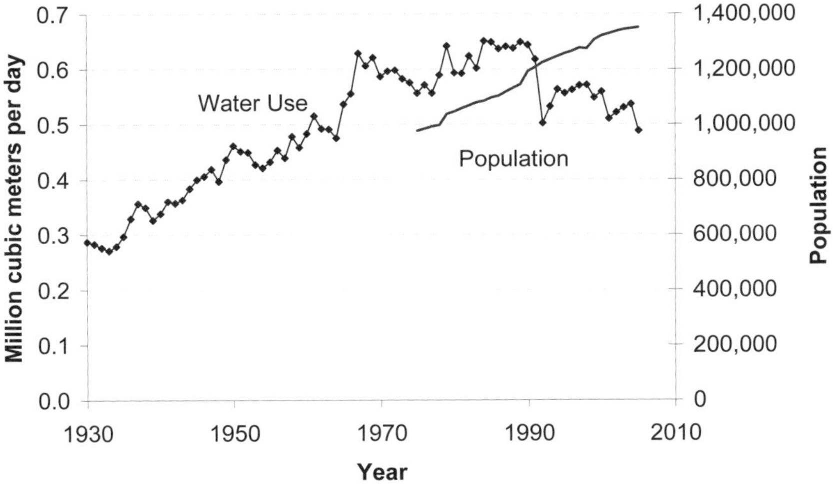

FIGURE 3-6. Water use in the city of Seattle. SOURCE: Data from Seattle Public Utilities (2007).

These trends in water demand have important implications for water planners, infrastructure development, and water policy. In particular, the traditional assumption by planners that economic and population growth lead to growth in total water withdrawals and necessary expansion of supply now appears to be incorrect or, at least, not inevitable. These changing trends also support the idea that improvements in water-use efficiency and shifts in economic structure can reduce resource use, even in an expanding economy (Gleick, 2003a), considering the ability to shift and reallocate water uses through the development of water markets, trades, or changes in water rights.

Future demands for water, however, may begin to rise again, and some regions of the United States may show far different trends for water use than the national average. For example, the growing emphasis on the production of biofuels is encouraging a shift to irrigation, even in regions like the Southeast where irrigation has not traditionally been used. Regional differences in water use are the result of variations in population dynamics, levels of effort to identify and reduce inefficiencies in water use, local water management practices, and other factors.

Review of U.S. Water Quality Challenges

Water quality is defined in response to variable concentrations of tens of thousands of natural and anthropogenic compounds, biologic material, many natural stable and radioactive elements, color, odor, and temperature. Even if an abundant quantity of water exists in a geographic region, water supply shortages may occur if the water quality is ill-suited to the specific societal demands. Water suitable for transportation of grain barges, cooling, or hydroelectric generation may not be appropriate for manufacture of semiconductor chips, public drinking supply, irrigation, or ecological uses. In addition, health studies are identifying new contaminants of concern, including methyl tertiary butyl ether (MTBE), perchlorate, pharmaceuticals, and more (NRC, 1996, 1999, 2005). Water-quality criteria (e.g., Safe Drinking Water Act amendments of 1996) are also evolving with time in response to epidemiological studies and technological advancements in analytical techniques and instrumentation that identify new environmental impacts and the presence of lower concentrations of solutes and biologic materials.

As large portions of the accessible nonsaline waters in the United States are fully allocated, some regions have little existing high-quality water available for expanding water supplies without negatively affecting environmental resources or other previously allocated water resources. Thus, water managers are increasingly turning to existing lower-quality water supplies, such as municipal and industrial wastewater, urban storm runoff, and saline water, such as seawater or brackish groundwater, to address new demands. Meanwhile, some existing water sources are facing increasing water-quality degradation due to nonpoint sources of pollution from land use and development, agricultural chemicals, and other practices (Kolpin et al., 1995; Snyder et al., 2000). Society has developed a wide range of engineering techniques and methods to modify water quality for specific uses. Desalination, for example, removes salts to permit the use of saline water sources for drinking water, irrigation, or industrial uses and has the ability to provide a very specific, tailored water quality from quite different source qualities. The wider use of membrane desalination technology may have the side benefit of helping to remove some emerging contaminants that otherwise are not removed by conventional treatment (Synder et al., 2007).

Water Use Projections for the Future

Designing and building major water infrastructure is time and capital intensive. As a result, water planners traditionally take a relatively long view by making projections of future demands. These projections produce expectations about future water use that in turn drive financial expenditures for water supply projects. Overestimates of supply needs can lead to unnecessary investment, environmental damage, and social and political risk. Underestimates of supply needs can lead to the failure to make timely investments, water supply shortages, and a reduction in system reliability. Water planners must balance these risks and benefits.

How can future water demands be predicted, given the uncertainties involved in looking into the future? Many projections of future U.S. water demands have been made over the past half-century (Brown, 1999; Thompson et al., 1971; USDA and USFS, 1989; U.S. Senate Select Committee on National Water Resources, 1961; Viessman, 1980; Wollman and Bonem, 1971; WRC, 1968, 1978). Review of the major studies that have been done reveals two noteworthy trends: overestimating total future water demand at the national level, often substantially, is the norm, not the exception; and, as tools and methods for making forecasts have improved, forecasts of future water needs have dropped (Gleick, 2003a).

Figure 3-7 shows more than 15 different U.S. water projections made before 2005 together with actual water withdrawals. As the figure shows, every projection made before 1995 substantially overestimated future water demands by assuming that use would continue to grow at, or even above, historical growth rates. The earliest projections routinely overestimated water demands because of their dependence on relatively simple assumptions about the relationship between water use, population, and

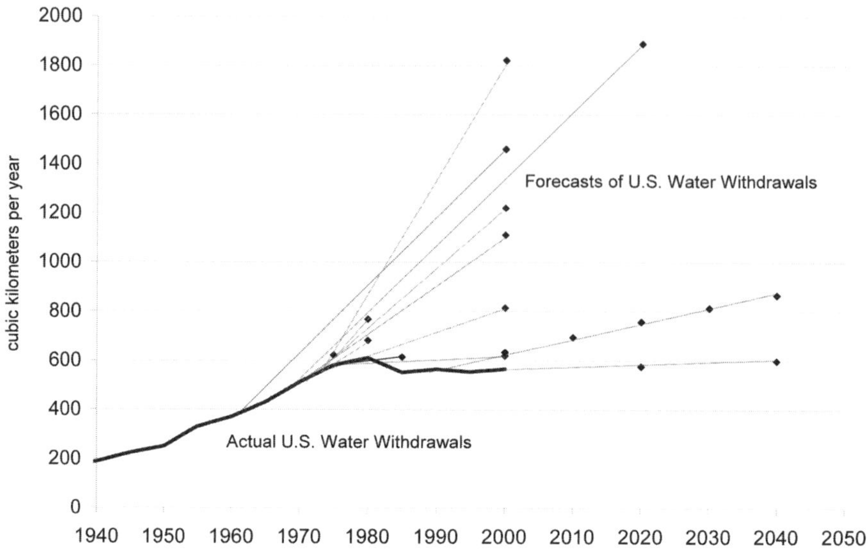

FIGURE 3-7. Projections of total U.S. water withdrawals since the 1960s. The points indicate projected withdrawals; the colored lines lead back to the date the projection was made. SOURCES: Water withdrawal data from MacKichan (1951, 1957), MacKichan and Kammerer (1961), Solley et al. (1998), and Hutson et al. (2003). Forecast data from Wollman and Bonem (1971), U.S. Senate Select Committee on National Water Resources (1961), WRC (1968, 1978), Viessman (1980), NWC (1971), USDA and USFS (1989), and Brown (1999).

industrial, commercial, and residential end-use intensity (e.g., water use per unit population, employment, revenue, or income). Agricultural water-use projections typically adopted constant end-use factors such as water per acre, which failed to incorporate either changes in water-use intensity of irrigation method, or changing cropping patterns. Most scenarios ignored water requirements for in-stream ecological needs, navigation, hydropower production, and recreation. And almost all of these forecasts showed dramatic increases in demand over time, sometimes to implausible levels of future use that led many observers to worry about water shortfalls and shortages. This history suggests that the traditional methods used by water-scenario developers are missing some critically important real-world dynamics (Gleick, 2003a).

Large-scale water-use projections have become increasingly sophisticated due to the growing capability of easily accessible computers to handle significant calculations and the growing availability of water-use data. Assessments that were conducted for continental areas or on a national basis are now being performed for watersheds on smaller and smaller temporal and spatial scales (Alcamo et al., 1997; Gleick, 1997).

The methods and tools used for forecasting and scenario analysis continue to rely heavily on traditional methods for making projections, but examples are emerging of new tools and new assumptions being used as planners reevaluate the driving factors behind changes in demands for water. Figure 3-8, for example, shows projections for Tampa Bay, Florida. This figure shows modest improvements in efficiency over time, but it still relies heavily on the assumption that per-capita water use will remain relatively unchanged while population grows into the future. Figure 3-9, however, shows forecasts produced by the City of Seattle, which has implemented a wide range of water management strategies to reduce sectoral and per-capita demand. As a result, Seattle's official water demand forecast now projects flat or declining water use through 2030 and small increases beyond that. Similarly, the State of California prepares a new water plan every 5 years, and the most recent version incorporates new estimates of significant water-use efficiency potential and evaluations of multiple scenarios. Even in the "current trends" scenario, the entire state is projected to use less water in 2030 than it uses at present, largely because of improvements in urban and agricultural water-use efficiency and changing land-use patterns (California Department of Water Resources, 2005).

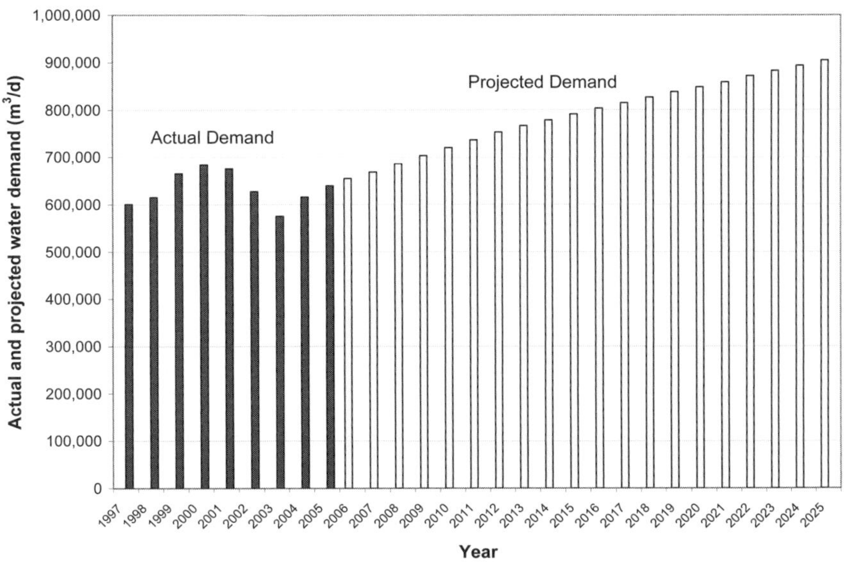

FIGURE 3-8. Actual and forecasted water demand for Tampa Bay, Florida. SOURCE: Jerry Maxwell, Tampa Bay Water, personal communication, 2007.

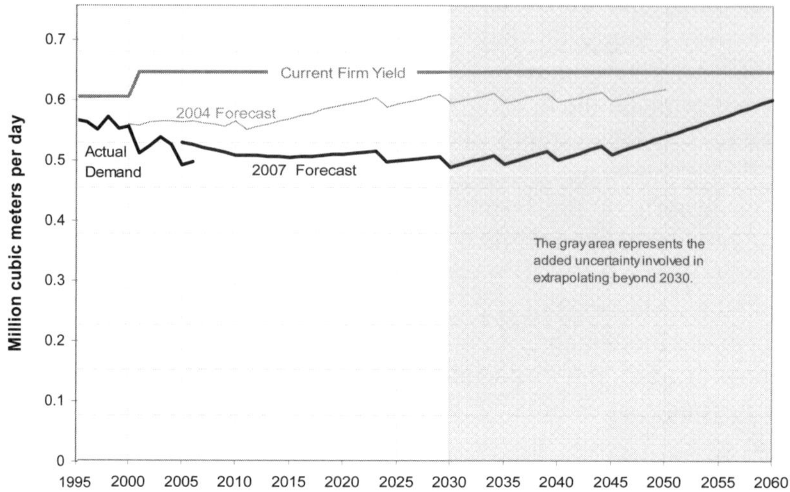

FIGURE 3-9. Actual demand, 2004 demand forecast, the most recent 2006 demand forecast, and estimated firm yield of water supply in the Seattle region. Note that the most recent demand forecast is well below the 2004 forecast, below average past demand through 2030, and remains below the firm yield of present supply through 2060.
SOURCE: Bruce Flory, Seattle Public Utilities, personal communication, 2008.

Although recent national water demand projections show relatively little increase in total water demand with time (Brown, 1999), some regions may show notably different trends than the national average. As mentioned previously, these regional differences result from variations in population growth, levels of effort to identify and reduce inefficiencies in water use, local water management practices, and other factors. Also, some regions may choose to reduce withdrawals from existing water supply sources for environmental reasons, thereby increasing the need for additional water sources while the total human demand remains unchanged (see Box 3-3).

OPTIONS FOR MAINTAINING WATER SUFFICIENCY

The definition of what is and is not a sufficient quantity and quality of water for a given purpose will vary with population, economic development, indigenous supplies of water, water quality considerations, and other factors. Perhaps the most significant feature of this concept is that it emphasizes the fact that past patterns of water use and past levels of

BOX 3-3
Desalination for Environmental Uses

While most of the discussion of desalination has focused on producing new water to supplement human uses, there is growing interest in the possibility of building desalination facilities that also directly or indirectly serve environmental needs. In some regions, human use of water has led to serious degradation of water quality and ecosystem habitat in rivers and streams. Efforts to restore that habitat rely, in part, on the ability of humans to reduce consumptive uses of water in order to return minimum flows to watersheds. In this context, some proposed desalination facilities could serve to replace, rather than supplement, existing sources of supply. For example, the California State Water Resources Control Board (SWRCB) ordered a reduction in diversions of water from the Carmel River in central California because of impacts on local threatened fish populations. The order (SWRCB, 1995) requires that any new supplies produced by the local water provider be used on a one-to-one basis to offset these diversions. One option under consideration is the construction of a desalination plant that would serve to satisfy the SWRCB order (Cooley et al., 2006).

sufficiency are not necessarily a guide to optimum patterns and levels of water use in the future. This comports with economic notions of supply and demand where the equilibrium quantities of water (and prices) are determined by the interaction of supply and demand. Inasmuch as the variables that affect supply and demand change over time, those equilibrium quantities (and prices) will change in response.

The most desirable quantities and prices for water supply are not always strictly determined by the elements of economic supply and demand, although those elements will always be important determinants of water sufficiency. Departures from strict supply and demand solutions may be caused by (1) the presence of technical external diseconomies;[1] (2) the public good attributes of some water services; (3) capital intensive industries such as airlines, natural gas, and water supply, where the dominance of capital or fixed costs prevents pricing the output according to its marginal cost; and (4) equity and welfare considerations. Despite the presence of one or more of these factors, the elements of supply and demand provide a useful framework for analyzing the various options available to maintain water sufficiency.

---

[1] External diseconomies are said to occur when some of the costs of an economic transaction fall on persons who were not parties to the transaction. Thus, for example, you may purchase an automobile. If in the manufacturing of that vehicle, some of the wastes of production are discharged into a river, then the manufacturer is externalizing the costs of waste disposal on people who use the river for recreation by degrading its quality. The recreational users are third parties to the decision to produce and sell the automobile and yet they are made to bear some of the costs. They are said to have borne the external diseconomies.

Historically, strategies for developing and managing water have tended to focus on supply-side options. These include the construction of storage and conveyance facilities that allow water to be captured and stored at wet times and places and transported for use at dry times and places. Most of the economical and environmentally acceptable surface-water storage opportunities in the United States have already been developed, and interest in storage has now shifted to the many opportunities for underground storage of water and the conjunctive or joint management of surface and groundwater. In addition, attention has tended to shift toward new forms of supply that convert previously unusable or unused sources of water into useful resources, such as wastewater reclamation and the desalination of brackish water and seawater. Although such technologies have tended to be very expensive, a combination of declining process costs and intensifying water scarcity is making them attractive options in some circumstances.

In recent decades, as supply options have become more constrained and expensive, attention has also been focused on demand management options. These options, which focus on tools to help reduce the amounts of water required to satisfy any given need, include (1) water metering and monitoring to inform people about actual water-use patterns, (2) education about the costs and benefits of water efficiency technologies and strategies, (3) pricing as a tool for encouraging efficient use and signaling consumers about the true costs of providing water, (4) water-use regulations and norms that encourage or require the use of xeriscaping and other efficiency technologies, (5) the elimination of subsidies that encourage inefficient water use or planting of low-valued, high-water-using agricultural crops, and (6) the development and adoption of water-saving irrigation technology and on-farm water-management practices.

There are also institutional options for addressing water scarcity, such as approaches for changing water allocations and rights. In some circumstances, water needed to sustain population and economic growth can be most economically obtained by reallocating water from relatively low-valued uses to higher-valued uses. This can be done through water markets or market-like mechanisms in which water can be exchanged, sold, or leased. Such market-like reallocations have been made recently from the Imperial Irrigation District in southeastern California and the San Diego County Water Authority. Other examples abound throughout the western United States (MacDonnell, 1990). These arrangements tend to be attractive in circumstances where the costs of supplementary supplies are relatively high and there are low-valued uses that make little sense to serve if higher-valued uses would otherwise go begging.

Water scarcity in some regions of the United States will certainly intensify over the coming decades. No one option or set of options is likely to be sufficient to manage this intensifying scarcity. Rather, the most successful management strategies will be developed by considering all of the available alternatives and emphasizing the development of those best adapted economically, technically, politically, and socially to the circumstances at hand. This comprehensive approach, sometimes called the "soft path," also acknowledges that different components of supply and demand management may differ in terms of the quality of the water, the cost of the water, and the reliability with which it is available (Gleick, 2002). When different tools are put together to meet particular management challenges, it is sometimes termed the "portfolio approach" (see Box 3-4).

BOX 3-4
Use of the Portfolio Approach to Water Management in San Diego County

The experience of the San Diego County Water Authority, a wholesaler of water, illustrates the use of multiple water-management approaches to address water demand. In 1991, the agency imported 95 percent of its supply through the Metropolitan Water District of Southern California. The remaining 5 percent came from local surface-water sources. Since that time, the San Diego County Water Authority has pursued a strategy of developing a broader portfolio of tools, diversifying its sources of supply, maximizing the efficient use of existing resources, and reducing its demands on imported supplies. Imported supplies tend to be vulnerable to large earthquakes and subject to other sources of uncertainty such as drought and legal disputes over water rights and water allocations. By 2005 the component of supply imported from the Metropolitan Water District accounted for only 79 percent of the total (see Figure 3-10). Conservation, the development and use of recycled water, an increase in the quantity of surface water supply, the development of some local groundwater, and water acquired via a market-like transfer from the Imperial Irrigation District were added to the portfolio.

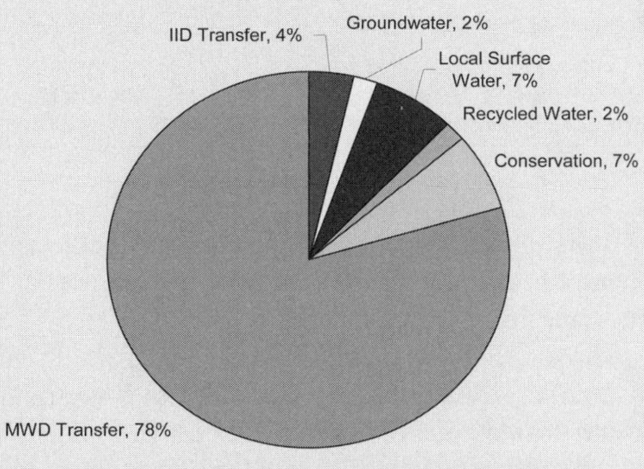

FIGURE 3-10: Water supply portfolio for San Diego in the year 2005.
SOURCE: Robert Yamada, San Diego County Water Authority, personal communication, 2006.

continued

As the population of the Authority's service area continues to grow, water availability will also have to grow if the new population is to be served. The San Diego County Water Authority estimates that it will need to increase supply by 185,022,276 m^3 between 2005 and 2020. It envisions that the component of water imported from the Metropolitan Water District will be reduced to half of what it was in 2005 and that conservation measures, including the elimination of percolation losses through canal lining, as well as the reliance on local sources and recycled water will all grow significantly (see Figure 3-11). In addition, it envisions that there would be a substantial component of desalinated seawater added to the portfolio. The target portfolio illustrates how the mix of sources of water availability can be diversified and can include a balance between old sources and new sources, demand management and supply augmentation, new supplies and recycled supplies, and the partial substitution of less remote supplies (from the Imperial Irrigation District) for more remote supplies (from the Metropolitan Water District).

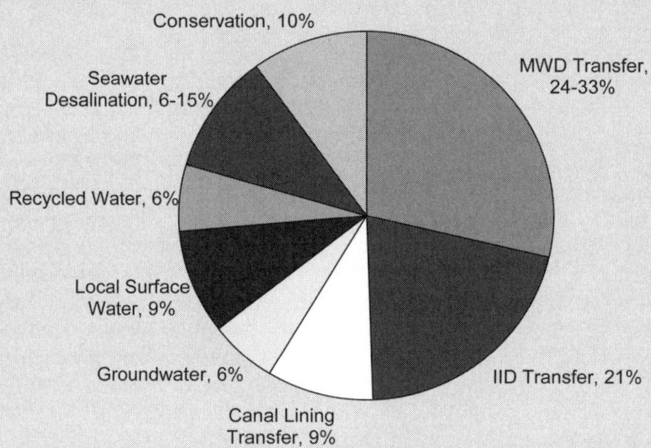

FIGURE 3-11: Water supply portfolio for San Diego in the year 2020.
SOURCE: Robert Yamada, San Diego County Water Authority, personal communication, 2006.

Desalination is an option to be considered, on balance with other alternatives, when planning for future water supply shortages. The addition of desalinated seawater to a water supply portfolio provides a source of "new" water, whose reliability is not linked to hydrologic variability (i.e., droughts). Of course, a water supply portfolio that is unbalanced in any direction can carry unexpected risks. Although the addition of desalination facilities can improve reliability under some circumstances, excessive reliance on desalination may have important energy implications in regions with constrained or unreliable energy supplies. Centralized desalination facilities may also carry security or seismic risks that require special attention.

CONCLUSIONS AND RECOMMENDATIONS

Long-term decisions about whether to pursue desalination facilities, other supply alternatives, efficiency improvements, reallocations and water transfers or trades, or other management policies hinge on many factors. While efforts to identify new, untapped sources of supply have dominated water policy for the past century, traditional sources of supply are increasingly expensive, unavailable, or controversial, raising new interest in desalination as a source of high-quality reliable supply. At the same time, however, new thinking about water demand and forecasting may be fundamentally reshaping the water debate by forcing a reevaluation of basic assumptions of ever-increasing water demand. Indeed, total water withdrawals in the United States have remained stable in recent years, despite growing populations and economy.

Changes in patterns of water consumption and withdrawals mean that past patterns of water use will not always be a reliable guide to the future. Satisfying future water demands requires integrating new factors and concepts into water planning, scenarios, and investment decisions. In particular, the assumption that water demands will inevitably parallel population and economic growth no longer appears to be correct, and rethinking this assumption offers new possibilities for water providers.

Most forecasts of future water demand have been unduly conservative and have failed to account for the fact that consumptive water use and withdrawals do not inevitably grow as population and economies grow. Total water use in the United States, including most regions and most sectors, has declined in recent decades. If these trends continue, pressure to identify new sources such as desalination may also decline, or at least be concentrated in particularly water-scarce regions or where users are able and willing to pay for high-quality, reliable new supply.

Desalination, using both brackish and seawater sources, is likely to have a niche in the water management portfolio of the future. The advantages of desalination, such as high reliability and insensitivity to natural hydrologic variability, are real but difficult to quantify and should be considered in any decisions among a "portfolio" of water management options. The significance of this niche, however, cannot be definitively determined at this time, because it will depend on a host of complicated and locally variable social, economic, environmental, and political factors, including societal preferences, the costs and reliability of alternative sources of supply, and the availability of cost-effective, environmentally sustainable concentrate management options. In the complete absence of these factors, the theoretical potential for desalination is effectively unlimited. Large quantities of inland brackish groundwater appear to be available (see Chapter 5) and, in coastal areas, ocean resources are essentially infinite in comparison to human demands. But, as with most resource questions, the theoretical potential and the practical potential are far different. The technological capabilities and the economic, social, and environmental constraints are discussed in detail in Chapters 4-7.

4

State of the Technology

Desalination technologies and their application have evolved substantially over the past 50 years. The five key elements of a desalination system (see Figure 4-1), for either brackish water or seawater desalination, are as follows:

1. Intakes—the structures used to extract source water and convey it to the process system;
2. Pretreatment—removal of suspended solids and control of biological growth, to prepare the source water for further processing;
3. Desalination—the process that removes dissolved solids, primarily salts and other inorganic constituents, from a water source;
4. Post-treatment—the addition of chemicals to the product water to prevent corrosion of downstream infrastructure piping; and
5. Concentrate management—the handling and disposal or reuse of waste residuals from the desalination system.[1]

Depending on the source water and the desalination technology used, specific elements may vary in their importance in the overall system. For example, inland brackish groundwater desalination facilities will use wells and pumps to bring the source water to the facility, and these systems may need little or no pretreatment. In contrast, seawater reverse osmosis (RO) desalination may use more elaborate intake structures, depending on the specific site conditions, and may require extensive pretreatment. The state of the technology for each of these elements and

[1] The broader term concentrate management (as opposed to concentrate disposal) is used throughout the report in recognition that waste management options are not limited only to disposal or discharge. However, it is worth noting that few economically viable concentrate reuse applications currently exist and that nearly all desalination concentrate in the United States is disposed rather than beneficially reused.

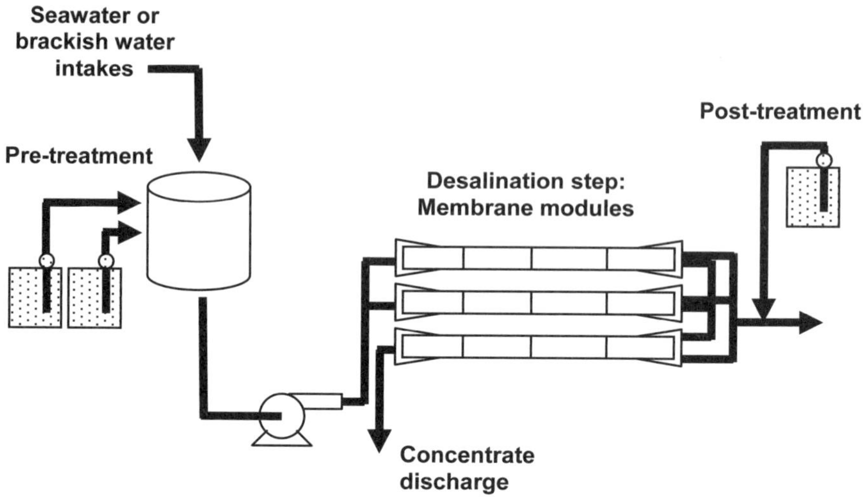

FIGURE 4-1. Key elements of a desalination system. Although shown here for a membrane-based system, these steps also describe the major components of non-membrane systems.
SOURCE: Modified from Buros et al. (1980).

their current technical barriers are discussed in this chapter. The focus of the chapter is on technologies that are commercially available or in the late stages of development, although some emerging technologies that are still in the early phases of research and development are discussed in boxes.

FEEDWATER INTAKE OPTIONS

Desalination facilities require a reliable supply of feedwater. Feedwater quantity and quality vary based on the specifics of the site and often determine the feasibility of siting a plant at a given location. Intake designs can affect feedwater quality and the environmental impacts of a desalination facility at a given site. Current technologies and issues with desalination intakes are discussed in this section.

Brackish water desalination facilities can utilize feedwater from surface water sources or wells. Inland desalination plants use intake technology that is no different from traditional water-treatment plants dependent on surface water or groundwater, and this technology is well developed. Therefore, these technologies will not be described in detail here. There are important environmental issues, however, associated with sustainable brackish groundwater withdrawals for inland systems that are discussed further in Chapter 5.

Seawater desalination intakes generally fall into one of two major categories—surface intakes located above the seafloor and subsurface intakes located beneath the seafloor or sandy beach. Surface intakes, often called open intakes, are in direct communication with the ocean or sea, where feedwater can be taken directly from the surface or below the surface through submerged intakes. Subsurface intakes take their feedwater from below the floor of the ocean using naturally occurring sand and geologic formations to provide filtration. Subsurface intakes can be horizontally drilled from central wells, slant drilled from onshore beaches, or excavated to create infiltration beds. Subsurface intake options can produce higher-quality feedwater and thereby reduce the pretreatment necessary for membrane desalination systems. In contrast, thermal seawater desalination systems require less pretreatment than RO and require only coarse screens to protect the process equipment. Thermal desalination plants commonly use surface (open) intakes.

Design engineering, equipment procurement, and construction spending on intakes and outfalls are estimated to total 5 to 7 percent of capital costs for RO and thermal desalination plants (GWI, 2006a). Other costs, such as monitoring and permitting, may add to the overall costs of the intakes. There are several factors affecting the final cost for constructing and operating an intake system; among these are the type of intake being used, the type of coastal conditions, and the distance from the intake to the plant itself. The following section focuses on the latest design and engineering options for coastal intakes.

Surface Intakes

Thermal seawater desalination and large seawater RO facilities (>38,000 m^3/day; >10 million gallons per day [MGD]) predominantly use open-water intakes. Screens are added to the intake structures to reduce the number of marine organisms taken in with the source water (referred to as entrainment; see also Chapter 5). Application of screen technology to the power industry has existed since the early twentieth century. Early screens included a front-end "trash rack" consisting of fixed bars to prevent large debris from entering the water intake system. Traveling screens are rotated and washed intermittently with a high-pressure wash. Alternative screen technology includes modified traveling screens with fish handling systems, fine-mesh screens, cylindrical wedge wire screens, fish net barriers, louvers, angled traveling screens, and velocity caps (California Water Desalination Task Force, 2003).

Impingement, defined as the pinning and trapping of fish or other larger organisms against the screens of the intake structures, can cause severe injury and death to organisms. State-of-the-art intake systems have been developed to greatly reduce impingement. For example, some intake screens can be backflushed with compressed air. These screens have no moving parts, operate with a very low velocity (to mitigate impingement), and are generally referred to as “passive screens.” A Ristroph screen is a modified traveling screen with water-filled lifting buckets that collect impinged organisms and transport them to a bypass, trough, or other protected area. Similarly, fish baskets consisting of framed screen panels can be attached to the vertical traveling screens. Fish that are removed are typically returned to the water via sluiceway or pipeline. Fine-mesh screens have a mesh size of 5 mm or less and are designed to exclude larger eggs, larvae, and juvenile fish from the intakes. Cylindrical wedge wire screens will exclude organisms larger than the nominal screen opening of 0.5 to 10 mm. Their cylindrical shape dissipates the velocity, allowing organisms to escape the flow field, although adequate countercurrent flow is needed to transport organisms away from the screen (California Water Desalination Task Force, 2003). An emerging screen technology that is currently in development to address impingement and entrainment is described in Box 4-1.

Louvers are a series of vertical panels placed perpendicular to the intake approach flow. They create a new velocity field that carries fish away from the intake and toward a fish bypass system. Louvers rely primarily on a fish’s ability to recognize the new flow field and swim away. They have been successful in reducing impingement but are not effective against the entrainment of eggs and larvae (California Water Desalination Task Force, 2003).

Shipboard seawater desalination approaches that situate the water treatment facility in the deeper ocean far from environmentally sensitive coastal areas could also reduce impingement and entrainment. One recent approach uses telescoping source water intakes to bypass the photic zone, where most marine organisms reside. This deeper water also contains fewer suspended solids, thus reducing the pretreatment required.

Subsurface Intakes

Coastal subsurface intakes include beach wells, radial wells, horizontal directionally drilled (also called slant-drilled) wells, and infiltration galleries. By taking advantage of the natural filtration provided by sediments, subsurface seawater intakes can reduce the amount of total organic carbon and total suspended solids, thereby reducing the pretreat-

BOX 4-1
Emerging Technologies to Reduce Impingement and/or Entrainment

A Marine Life Exclusion System has been designed to reduce impingement as well as entrainment. This water-permeable barrier (see Figure 4-2) is spread around the intake structure, preventing aquatic organisms from approaching the water intake point. The curtain is either suspended by flotation billets and anchored in place, or integrated into existing shoreline intake structures. Sealed against the seafloor and shoreline structures, it completely surrounds the intake structure, preventing targeted planktonic and neustonic organisms from entering the system. Because the surface area of the curtain is large compared to an intake screen, the water velocity through the curtain is up to 98 percent less than the velocity near the intake structure. Low water velocity enables even small fish larvae to drift away from the boom. This technology has primarily been used in riverine environments, although it is currently being tested in marine settings to examine its durability, susceptibility to fouling, and cleaning requirements (McCusker et al., 2007; Mirant Lovett, 2006; San Francisco Bay Conservation and Development Commission, 2005).

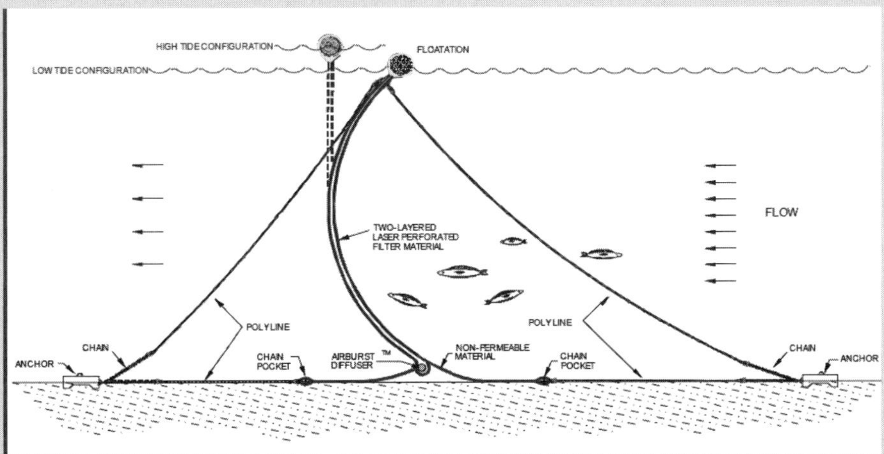

FIGURE 4-2. Curtain designed to reduce intake velocities and minimize impingement and entrainment.
SOURCE: McCusker et al. (2007)

ment required for membrane-based desalination systems and lowering the associated operations and maintenance costs. Pumping from subsurface intakes may also under some conditions dilute the seawater with less saline groundwater, thereby reducing the total dissolved solids (TDS) in the intake water.

Vertically drilled beach wells are typically used for small (<19,000 m^3/day; <5 MGD) systems where the local hydrogeology (e.g., aquifer transmissivity) will permit it. Beach wells have been used effectively in the Caribbean and Mediterranean and are the intakes of choice for proposed plants in Hawaii. They pose minimal environmental concerns because benthic communities remain undisturbed and entrainment and impingement of marine organisms are eliminated. In most cases, no further pretreatment is needed. One of the potential disadvantages of beach wells is that deep wells may result in lower water temperature and thus higher viscosity; hence, higher pressure (and increased energy) will be required to pump the water through the RO membranes. Care should also be taken to ensure that the source water withdrawals do not cause deleterious effects on local aquifers.

Horizontal directionally drilled (or slant-drilled) wells, shown in Figure 4-3, are increasingly being considered for use in large seawater desalination facilities. Although more expensive to construct than beach wells, they can minimize shoreline structures. Slant-drilled wells are under study in Dana Point, California, for example. They are also currently in use at several seawater RO plants in Spain, including the facility at San Pedro del Pinatar, which has a capacity of over 170,000 m^3/day (Peters et al., 2006).

An alternative approach recently used at the Fukuoka, Japan plant is a seabed infiltration gallery (Figure 4-4). This intake system requires isolating a section of beach so that sand can be removed to the desired depth. Varying grades of small rock and gravel are placed into the excavation and perforated pipes are installed to convey source water to the plant. The rock and gravel are covered with the same sand material that was excavated from the beach, before the ocean is allowed to resume its

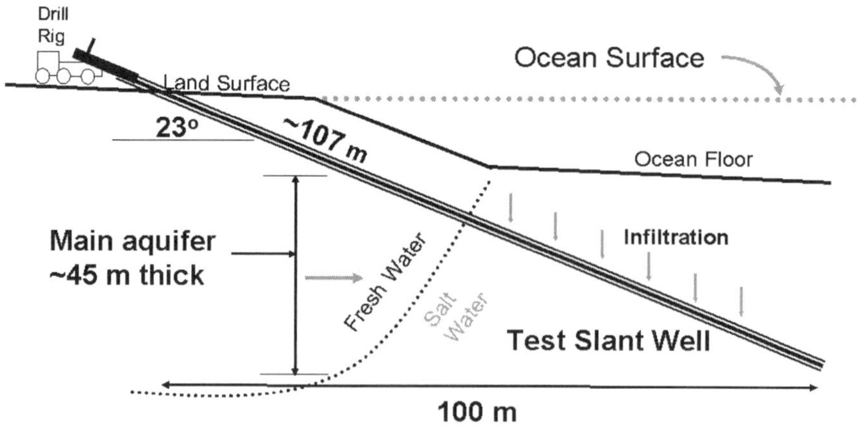

FIGURE 4-3. Slant-drilled well concept.
SOURCE: Adapted from Richard Bell (2006).

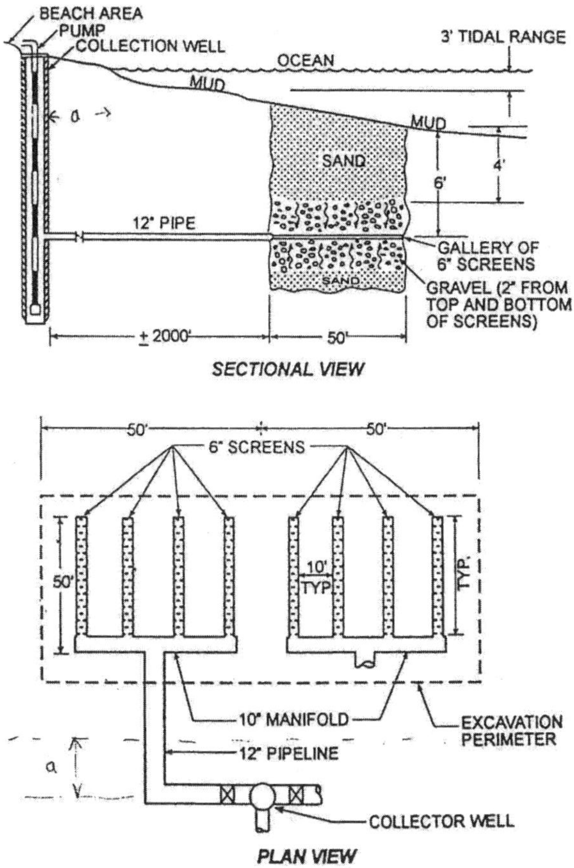

FIGURE 4-4. Seabed filtration system.
SOURCE: Missimer (1994).

normal shoreline. The benefits of this approach are that the source water is nearly all seawater and not diluted by freshwater aquifer contributions and that it greatly mitigates entrainment and impingement (Wright and Missimer, 1997). Although proven technically successful in Japan, theenvironmental impacts will need to be better understood and, if needed, mitigated.

PRETREATMENT

Pretreatment is generally required for all desalination processes. Pretreatment ensures that constituents in the source water do not reduce the performance of the desalination facility. Thermal processes require pretreatment to avoid scaling and to control corrosive constituents of the source water. Some removal of sand or gritlike suspended solids may also be necessary to avoid pipe erosion. In membrane desalination, pretreatment involves these considerations as well as further pretreatment to remove suspended solids of both biological and mineral origin to avoid membrane fouling. Biological growth may need to be inhibited by a disinfectant or biocide.

Pretreatment is a critical step in seawater and brackish water membrane desalination systems that utilize feedwater from surface water sources, because the suspended and colloidal particles, organisms, and natural organic matter need to be removed before the feedwater reaches the membranes. Indeed, proper pretreatment of feedwater is the most important factor in the successful operation of an RO plant, and pilot testing of the pretreatment process is a critical part of plant design.

Brackish water desalination systems that treat groundwater require very minimal, if any, pretreatment to remove particulates because the water typically contains very low concentrations of suspended solids and organic matter. Nevertheless, brackish groundwater may require pretreatment to remove selected constituents such as dissolved iron, manganese, and sulfides, which, if oxidized, create particulates that can foul RO membranes (USBR, 2003).

The quality of source water available at a particular site will also affect the extent of pretreatment needed for membrane desalination. Source water quality will depend on local site factors such as source water depth, turbidity, boat traffic, oil contamination, nearby outfalls, wind conditions, tides, and the influence of runoff. As discussed previously, subsurface seawater intakes, aquatic filter barriers, and deep ocean water intakes can greatly reduce the need for pretreatment. Due to permitting regulations and available land, however, desalination plants cannot always be sited where they will have the lowest pretreatment costs. Furthermore, because the United States employs more rigorous accounting of environmental costs, siting options in the United States may lead to greater pretreatment and greater pretreatment residuals handling needs compared to global standards of practice. The most common pretreatment processes are discussed below.

Scaling and Corrosion Control

Scaling is caused by the precipitation of minerals, such as calcium carbonate, from solution. Calcium sulfate scaling can be controlled via temperature control or through pretreatment by nanofiltration to remove the calcium ions. Acidification of the feedwater can prevent calcium carbonate or magnesium hydroxide formation and scaling. Finally, the use of chemical antiscalants such as sodium hexametaphosphate or polymeric acids can sequester the cations that can lead to scaling problems (Table 4-1).

Corrosion can be reduced by removing corrosive gases in pretreatment. Carbon dioxide can be controlled through acidification, and oxygen can be controlled with an oxygen scavenger such as sodium bisulfate or ferrous sulfate. Alternatively, corrosion can be controlled in some systems through the formation of a protective film within the system by adding zinc orthophosphate (Table 4-1).

Conventional Solids Removal Methods

Conventional solids removal methods such as coagulation and sedimentation followed by media filtration are still the predominant pretreatment processes for seawater RO. Chemicals such as ferric chloride or polyelectrolytes are added to enhance the coagulation of suspended solids prior to settling and filtration (Table 4-1). Traditional gravity flow filtration has been successfully used at many seawater RO plants around the world. At Point Lisas, Trinidad, gravity filters with greater-than-normal depth proved to be successful in pretreating seawater that encounters severe spikes in turbidity due to the intake location in a ship turning basin (Jacangelo and Grounds, 2004). These are mature technologies, although novel approaches to conventional filtration continue to be examined. For example, at Tampa Bay, an upflow dual sand process was installed that had previously only been used for industrial and wastewater applications (see Box 4-2). Nevertheless, there is a need to improve the quality and stability of influent to RO membranes; thus, other pretreatment options continue to emerge.

Microfiltration and Ultrafiltration

Microfiltration (MF) and ultrafiltration (UF) membranes are increasingly being used in the pretreatment processes for membrane desalination. Water molecules and salts are free to pass through, and water is pushed (or pulled) through the membrane at very low pressures. Particles larger than the membrane pore size (0.03-10 μm for MF and 0.002-0.1 μm for UF) are removed. Membranes are commercially available in flat-sheet, tubular, hollow-fiber, and spirally wound configurations. Among the benefits of MF/UF pretreatment compared to conventional pretreat-

TABLE 4-1 Reported Dosing Concentrations of Pretreatment Chemical Additives in Reverse Osmosis and Multistage Flash Desalination

Chemical Additive	Reported Dosing (mg/L)	References
REVERSE OSMOSIS DESALINATION		
Biocide		
Chlorine	0.5-6	Abart, 1993; Redondo and Lomax, 1997; Morton et al., 1996; Woodward Clyde Consultants, 1991
Chlorine removal		
Sodium bisulfite	3-19	Morton et al., 1996; Redondo and Lomax, 1997; Woodward Clyde Consultants, 1991
Coagulants		
Ferric chloride	0.8-25	Baig and Kutbi, 1998; Woodward Clyde Consultants, 1991
Polyelectrolyte	0.2-4	Ebrahim et al., 1995; DuPont, 1994; Hussain and Ahmed, 1998
Antiscalants		
Sulfuric acid	6.6-100	Al-Shammiri et al., 2000; Morton et al., 1996; Al-Ahmad and Aleem, 1993
Sodium hexametaphosphate (SHMP)	2-10	Al-Ahmad and Aleem, 1993; Al-Shammiri et al,. 2000 FilmTec, 2000
Polyacrylic acid	2.9	Woodward Cycle Consultants, 1991
Phosphonate	1.4	Al-Shammiri et al., 2000
MULTISTAGE FLASH DISTILLATION		
Biocide		
Chlorine	0.25-4	Iman et al., 2000; Shams El Din and Makkawi, 1998; Khordagui, 1992; Abdel-Jawad and Al-Tabtabaei, 1999
Hypochlorite	2	Burashid, 1992
Antiscalants		
Polyphosphate	2.2-2.5	Hamed et al., 2000; Abdel-Jawad and Al-Tabtabaei, 1999
Polycarboxylic acid	1.5-2	Hamed et al., 1999
Polyphosphonate	1-3	Hamed et al., 1999, 2000
Antifoaming agents		
Polypropylene glycol	0.035-0.15	Imam et al., 2000
Corrosion control		
Sodium bisulfite	Not given	Imam et al., 2000
Ferrous sulfate	1 - 3	Shams El Din and Makkawi, 1998
B-ethyl phenyl ketocyclohexylamino hydrochloride	25	Andijani et al., 2000

NOTE: The types and concentrations of pretreatment chemicals vary with plant design and source water conditions. Thus, some or all of these chemicals may not be used at all, or they may only be used intermittently. Some added chemicals can be recovered or removed (e.g., chlorine).
SOURCE: Adapted from Lattemann and Höpner (2003).

BOX 4-2
Pretreatment Changes at the Tampa Bay Seawater Desalination Plant

The largest seawater desalination plant in the United States (95,000 m^3/day or 25 MGD) is located in the Tampa Bay region of Florida. The Tampa Bay Seawater Desalination project obtains its source water from a "once-through" cooling system at the Tampa Electric Company (TECO) Big Bend Power Station, which withdraws its cooling water from Tampa Bay. After coming online in March 2003, the plant experienced performance problems that significantly stemmed from an inadequate pretreatment system. The desalination plant did not include any additional screens beyond those that existed within the power plant, and the upflow sand filter system was inadequate to produce pretreated seawater adequate to sustain reverse osmosis process operation. Although the warmer water from the power plant requires less energy for the desalination process, higher water temperature introduces greater potential for biological growth and pretreatment challenges. The original pretreatment process (Figure 4-5) resulted in severe premature fouling of the cartridge filters after 3-4 days of operation and premature membrane cleaning immediately following a 2-week acceptance test.

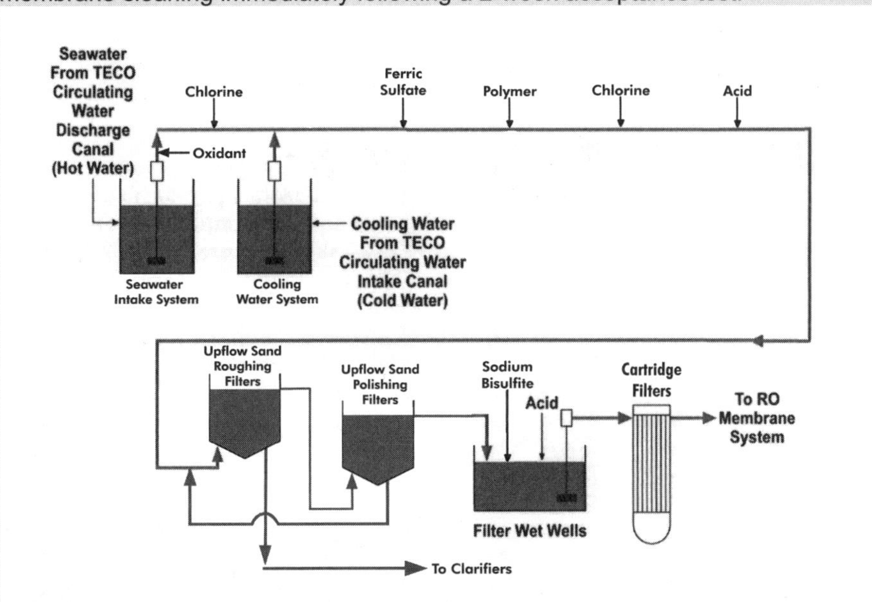

FIGURE 4-5. Original pretreatment design for the Tampa Bay Water Desalination Plant.
SOURCE: Adapted from figure courtesy of Tampa Bay Water.

The new design (Figure 4-6) is a much more robust treatment process that includes significant changes to the head works of the plant. Changes include the addition of 1/16-inch traveling screens for debris removal, ferric chloride as a coagulant, the use of chlorine dioxide for biological growth control, mechanical chemical mixing and coagulation basins to achieve better floc formation, and extended flocculation zones to aggregate and settle suspended particles up

continued

stream of the sand filters. Water distribution to the individual sand filters is now individually piped, valved, and metered to allow remote control of specified flow to each filter. The upflow sand filters were converted from a dual-stage, roughing and polishing configuration to a single-stage system, which reduced the hydraulic loading rate to each filter. Diatomaceous earth filters were added as the polishing step after the upflow sand filters to complete the particulate removal process immediately prior to the 5-μm cartridge filters. Development of a robust pretreatment process will reduce the fouling rate of the cartridge filters and cleaning requirements of the reverse osmosis membranes. Reducing the membrane cleaning frequency will extend the useful life of the roughly 10,000 reverse osmosis membrane modules at the plant, protecting the membrane installation cost of approximately $6 million.

The Tampa Bay pretreatment system depicted in Figure 4-6, the first of its kind for seawater desalination, is not a proven pretreatment technology. Because pretreatment is a key step to successful operation of seawater desalination plants, it is critical that new pretreatment approaches be tested systematically before implementation. The capital cost of the Tampa Bay pretreatment system is estimated to be approximately 30 percent ($32 million) of the total engineering and construction cost ($108 million). The pretreatment operating costs, including power, chemicals, and diatomaceous earth, is $0.12 per cubic meter ($0.44 per thousand gallons) (Joe Dysard, R.W. Beck, personal communication, 2007).

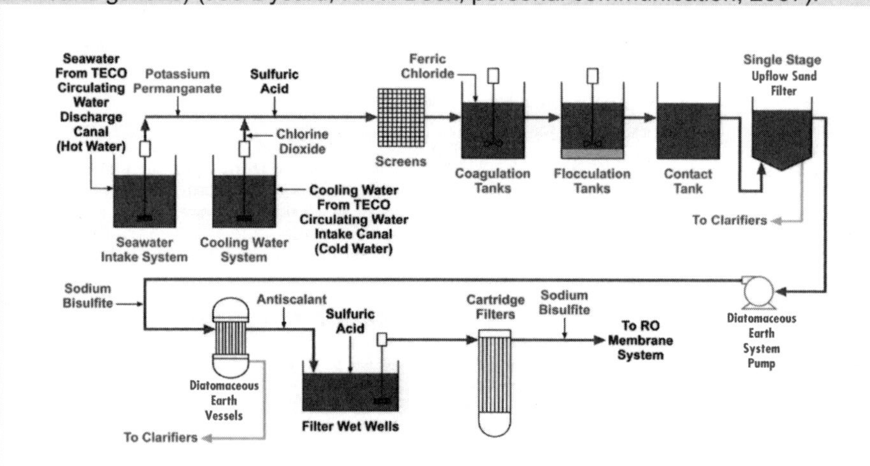

FIGURE 4-6. Modified pretreatment process at Tampa Bay Desalination Plant. SOURCE: Adapted from figure courtesy of Tampa Bay Water.

ment technologies are (1) production of feedwater to the RO system of constant and high quality regardless of source water fluctuations; (2) reduced RO fouling, which results in less cleaning and longer membrane life; (3) smaller footprint; and (4) lower consumption of chemicals. Potential disadvantages include higher costs and negative environmental impacts of concentrate from these membranes.

MF and UF membrane processes have been developed and piloted for application in seawater pretreatment. Several studies highlight the benefits of using MF or UF as pretreatment for RO desalination (e.g.,

Allam et al., 2003; Galloway and Mahoney, 2004; Latorre, 2001; Pearce, 2007; Taniguchi et al., 1995; Xu et al., 2007, Zhang et al., 2006).

Biofouling Control

Chlorine or hypochlorite have been the standard oxidants used for biofouling control, but thin-film composite polyamide membranes commonly used in RO desalination cannot tolerate oxidants like chlorine, and the chlorine needs to be removed in pretreatment by addition of a reducing agent such as sodium bisulfite. Sodium bisulfite and copper sulfate can also be used as biocides in membrane systems.

Ultraviolet (UV) and ozone treatment are being considered potential replacements for chlorine-based biological growth control of RO feedwater. Both UV and ozone have merit and have been successfully used in small to midsized drinking water and water reuse applications globally. UV will not cause problems with oxidant-sensitive membranes. Ozone is a much more effective disinfectant, but it poses a problem to the oxidant-sensitive RO membranes (Cotruvo, 2005; Glater et al., 1983).

Assessment of Pretreatment Technology

Conventional pretreatment for RO (i.e., coagulation and sand filtration) is a mature technology and has been proven effective for seawater desalination plants with surface water intakes. MF and UF pretreatment systems for RO are likely to gain more popularity due to the superior quality of water that such systems can produce, although their robustness under such conditions remains unproven. High-quality pretreatment will be of greater importance with the advent of "high-flux" (high-permeability) RO membranes, as the propensity for fouling is significantly increased at higher water fluxes. To realize the benefits of high-flux RO membranes, more effective pretreatment systems will be needed.

DESALINATION PROCESSES

The desalination process represents the step in which dissolved solutes are substantially removed from the feedwater to create the desired product water. A number of technologies exist to accomplish this objective, including the more commonly used membrane, thermal, and ion-exchange processes.

Membranes can be designed to selectively permit or prohibit the passage of certain ions, including salts. Membranes play an important role in the separation of salts in the natural processes of dialysis and osmosis, and this natural principle has been adapted in two commercially important desalting processes: electrodialysis (ED) and RO. In recent years, significant advances in RO technology have been achieved and, globally, more new membrane desalination capacity is now added annually than distillation capacity (Figure 4-7). Currently, membrane processes, including RO, nanofiltration (NF), ED, and electrodialysis reversal (EDR), account for 56 percent of the online capacity for desalination worldwide.

The basic concept of thermal distillation is to heat a saline solution to generate water vapor. If this vapor is directed toward a cool surface, it can be condensed to liquid water containing very little of the original salt. Water will boil under atmospheric pressure at 100°C, but thermal processes can also be designed to boil water in a series of vessels operating at successively lower temperatures and pressures. At one-quarter of normal atmospheric pressure, water will boil at 65°C, and it will boil at only 45°C if the pressure is decreased to one-tenth normal

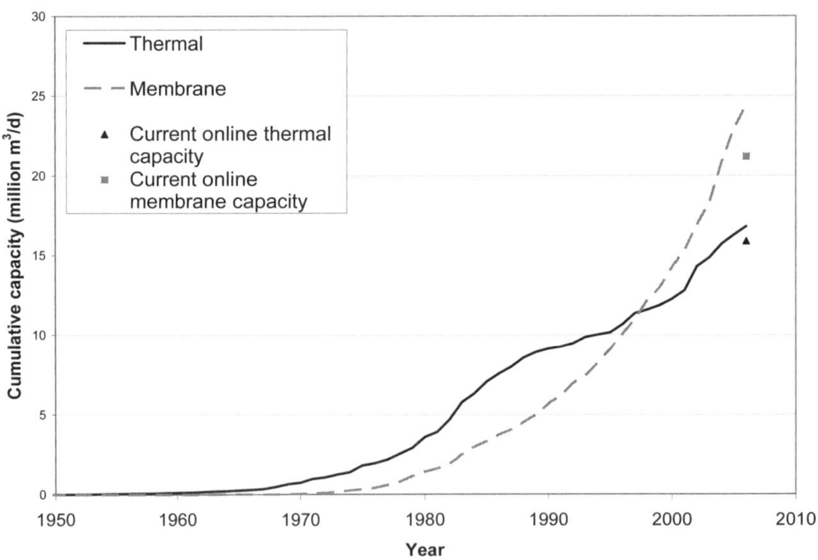

FIGURE 4-7. Cumulative global capacity of installed desalination plants for thermal and membrane technology. Thermal technology includes MED, MSF, and MVC. Membrane technology includes RO, NF, ED, and EDR. Points reflect current online (or presumed online) capacity of both technologies.
SOURCE: GWI (2006b).

atmospheric pressure. The concept of distilling water with a vessel operating at a reduced pressure has been used for well over a century.Thermal processes, such multistage flash (MSF), multiple effect distillation (MED), and mechanical vapor compression (MVC), account for 43 percent of the online capacity for desalination worldwide.

Water can be desalted through many other processes, including small-scale ion-exchange resins and hybrid processes. None, however, has achieved the commercial success of membranes or thermal distillation. Together these other processes account for less than 1 percent of total desalination capacity worldwide (GWI, 2006b).

There is no single "best" method of desalination. Globally, thermal and membrane technologies are both used widely for seawater desalination. Both processes require energy to effect separation of salts, and various energy sources can be used. Brackish water is typically desalinated using RO, NF, or ED. Significant numbers of smaller plants have also been built that use vapor compression and other methods (GWI, 2006b; USAID, 1980). The dramatic growth in membrane and thermal desalination processes over the past four decades is shown in Figure 4-7.

The objective of this section is to present the state of the technology of currently applied desalination processes that might have potential for addressing water needs in the United States. Energy use, performance, theoretical limitations, and current technical issues that need to be improved will be addressed in this section. The specific energy for desalination (i.e., the energy needed to produce a unit volume of fresh water from a saline feedwater) is an important consideration when comparing the various desalination technologies in this section. Potential improvements of these processes are discussed, considering that several of the technologies are mature with respect to approaching the value of the theoretical minimum energy for desalination (Box 4-3).

Membrane Desalination Processes

Commercially available membrane technologies for desalination include RO, NF, and ED or EDR. RO and NF are classified as pressure-driven membrane processes, whereas ED and EDR are electrically driven (see Box 4-4). Membrane technologies can be used not only for desalting brackish water and seawater sources but also for treating wastewater in reuse and recycling applications, because of their ability to provide removal of nonsalinity contaminants (e.g., organic contaminants, bacteria, and viruses; see Figure 4-9). Typically, 35 to 60 percent of the seawater fed into a membrane process is recovered as product water. For brackish water desalination, water recovery can range from 50 to 90 percent, de-

BOX 4-3
Theoretical Minimum Work (Energy) for Desalination

Desalination requires an intrinsic minimum *available* energy. Available energy, often called *exergy*, is mechanical, electrical, or any other energy, which in practice can be nearly completely converted into mechanical work (Spiegler and El-Sayed, 1994). Any desalination process will be most efficient if it involves a *reversible* thermodynamic process. In any reversible desalination process, the same energy is needed to desalt water, and this energy is independent of the technology or device employed and the exact mechanism of desalination. Thus, all desalination systems share a theoretical minimum work (available energy) that is independent of the system used.

There are several approaches for calculating the theoretical minimum energy for desalination. Stoughton and Lietzke (1965) based their analysis on the fact that the free energy change involved in removing a small amount of pure water from a mixture of water and salt is equal in magnitude but opposite in sign to that for adding the same amount of pure water to the mixture. They have further shown that the free energy change is equal to the minimum energy needed for desalination. The minimum energy for desalination for any percent of water recovery (up to saturation of salt) and a wide range of temperatures (from 25 to 200°C) was calculated. For zero percent recovery, that is, the removal of a relatively small amount of water from a very large amount of seawater, the calculated theoretical minimum energy for desalination was 0.70 kWh/m^3 of freshwater produced.

Another approach which resulted in similar results for the theoretical minimum work for desalination was presented by Spiegler and El-Sayed (1994). In this approach, compartments containing saltwater and freshwater are considered. The compartments are separated by a semipermeable membrane that permits water but not salt to pass through it. At equilibrium, hydraulic pressure must be exerted on the salt solution to prevent spontaneous transport of water from the freshwater to the salt solution. This hydraulic pressure is equal to the osmotic pressure, Π_{os}. To separate water from the salt solution, the pressure on the salt solution is increased under reversible conditions, so that the applied pressure exceeds the osmotic pressure by only an infinitesimal amount. A small volume of pure water, dV, is passed through the membrane, but because of the increase in the salt solution concentration and osmotic pressure, there is a need to increase the applied pressure again to ensure transport of water from the salt solution to the freshwater. This process continues until the desired amount of water has been removed from the salt solution (i.e., when a desired water recovery is attained).

The differential work, dW, needed for obtaining a differential amount of fresh water, dV, is given by (Spiegler and El-Sayed, 1994)

$$dW = \Pi_{os} dV \qquad (1)$$

Denoting the initially volume of the salt solution as V and the final volume as V, the total work per volume of freshwater produced, W, is

continued

$$W = \frac{1}{V_1 - V_2} \int_{V_1}^{V_2} \Pi_{os} dV \qquad (2)$$

The osmotic pressure is a function of the activity of water, and it decreases with increasing salinity of the salt solution. These activity coefficients can be determined from the ratio of the vapor pressure above the salt solution to the vapor pressure of pure water. Following this analysis, Spiegler and El-Sayed (1994) showed that the theoretical minimum work for zero recovery at 25°C is 0.70 kWh/m^3, similar to the value reported by Stoughton and Lietzke (1965) as discussed earlier. Furthermore, the theoretical minimum energy increases for salt solutions of higher temperatures. For instance, the theoretical minimum energy for zero recovery is 0.72, 0.82, and 0.87 kWh/m^3 for seawater temperatures of 50, 75, and 100°C, respectively (Stoughton and Lietzke, 1965).

It was further shown that, for any water recovery, the minimum work is related to the theoretical minimum work for zero recovery, W_0, via (Spiegler and El-Sayed, 1994):

$$W = W_0 \frac{V_1}{V_1 - V_2} \ln \frac{V_1}{V_2} \qquad (3)$$

It is useful to express the minimum work for desalination in terms of the water recovery, r, noting that $r = (V_1 - V_2)/V_1$, as follows:

$$W = \frac{W_0}{r} \ln \frac{1}{1-r} \qquad (4)$$

The minimum work for desalination as a function of water recovery, calculated by the preceding analysis, is presented in Figure 4-8. As seen, the theoretical minimum energy for desalination increases dramatically at high water recoveries.

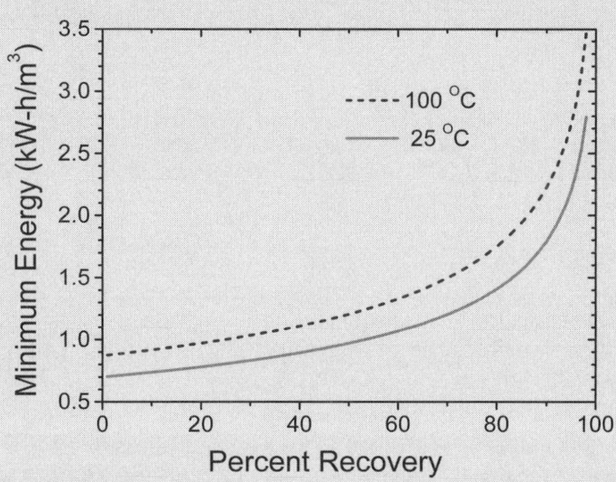

FIGURE 4-8. Minimum work (energy) for desalination of seawater per cubic meter of produced water. Calculations are based on a seawater solution (3.45 wt% of salts) and for temperatures of 25 and 100°C.

pending on initial salinity and the presence of sparingly soluble salts and silica, although recovery is typically between 60 and 85 percent (Sethi et al., 2006a). The remaining reject salt solution becomes more concentrated and must be disposed. For both brackish water and seawater, membrane processes can reduce salinity in the product water to levels less than 500 ppm TDS.

Reverse Osmosis

The RO process uses semipermeable membranes and a driving force of hydraulic pressure, in the range of about 1,000 to 8,300 kilopascals (kPa) (150 to 1200 pounds per square inch [psi]; 10 to 83 bar), to remove

BOX 4-4
Membrane Systems

The major membrane types that can be used for desalination and/or pretreatment are the folllowing:

Electrodialysis (ED) is an electrochemical separation process in which ions are transferred through ion-exchange membranes by a direct current voltage, leaving desalinated water as the product.

Reverse osmosis (RO) membranes desalinate both brackish water and seawater by applied pressure using a solution/diffusion mechanism whereby the water dissolves into and diffuses through the nonporous membrane, leaving the majority of the salts behind in the concentrate. RO membranes are also capable of removing some larger organic contaminants. Small uncharged species can pass through the membrane.

Nanofiltration (NF) membranes are used for water softening, organics and sulfate removal, and some removal of viruses. Pressure-driven removal is by combined particle size-based sieving and solution/diffusion. Pores in NF membranes are usually smaller than 0.001 μm and a molecular weight cutoff (MWCO) of 1,000 to 10,000 daltons.

Ultrafiltration (UF) membranes are used for removal of contaminants that affect color, high-weight dissolved organic compounds, bacteria, and some viruses. UF membranes operate via a pressure-driven size-based sieving mechanism through a membrane with pores in the range of 0.002 to 0.1 μm with an MWCO of 10,000 to 100,000 daltons.

Microfiltration (MF) membranes are used to reduce turbidity and remove suspended particles, algae, and bacteria. MF membranes operate via a sieving mechanism under a lower pressure than either UF or NF membranes, through membrane pores of 0.03 to 10 μm and an MWCO of greater than 100,000 daltons.

continued

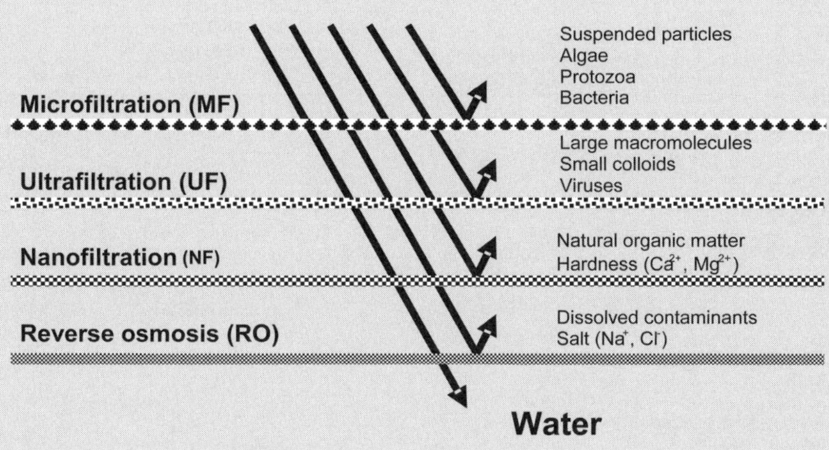

FIGURE 4-9. Substances and contaminants nominally removed by pressure-driven membrane processes. Modified from Trussell and Trussell (2005). SOURCES: NRC (2004b), Cooley et al. (2006), Sedlak and Pinkston (2001), Heberer et al. (2001), AWWA (1999), AWWARF et al. (1996), and NRC (1997).

dissolved solids from brackish water or seawater. The process can be described as solution/diffusion controlled, because the ions move through RO membranes via the process of diffusion (Lonsdale et al., 1965). Salts do permeate the membrane but at permeabilities that are orders of magnitude lower than that of water; thus, the majority of dissolved salts are removed by the process. RO can also remove synthetic organic chemicals and disinfection by-product precursors. However, dissolved gases such as hydrogen sulfide and carbon dioxide and some pesticides or low-molecular-weight organics pass through RO membranes.

The specific energy for RO desalination varies with the system used, the operational conditions (e.g., flux, recovery), and the quality of feedwater to the RO system. For seawater RO, the specific energy usage is typically about 3-7 kWh/m^3 with energy recovery devices (Alonitis et al., 2003; Miller, 2003; see Table 4-2). For brackish water RO, energy usage is comparatively lower, about 0.5-3 kWh/m^3, because the energy required for desalination is proportional to the feedwater salinity (Sethi et al., 2006b; see Table 4-3 and Figure 4-10). Energy usage values should be taken cautiously because the "system" for which desalination energy use is calculated and reported (i.e., basic RO process only, or including other ancillary equipment or processes) varies in the literature.

RO membrane formulations include cellulose acetates, polyamides,

TABLE 4-2 Comparison of Predominant Seawater Desalination Processes

	Seawater RO	MSF	MED (with TVC)	MVC
Operating temperature (°C)	<45	<120	<70	<70
Pretreatment requirement	High	Low	Low	Very low
Main energy form	Mechanical (electrical) energy	Steam (heat)	Steam (heat and pressure)	Mechanical (electrical) energy
Heat consumption (kJ/kg)	NA	250-330	145-390	NA
Electrical energy use (kWh/m^3)	2.5-7	3-5	1.5-2.5	8-15
Current, typical single train capacity (m^3/d) [a]	<20,000	<76,000	<36,000	<3,000
Product water quality (TDS mg/L)	200-500[b]	< 10	< 10	< 10
Typical water recovery	35-50%	35-45%[c]	35-45%[c]	23-41%[c]
Reliablility	Moderate	Very high	Very high	High

[a] For the purpose of this table, a train is considered a process subsystem which includes the high-pressure pump, the membrane array(s), energy recovery devices, and associated instrumentation/control. However, larger facilities may group pumps, membranes, and energy recovery into process or pressure centers to lower capital costs and improve operating costs.

[b] Product water quality for RO is a design variable. Each pass through an RO plant typically removes 99 to 99.5 percent of dissolves salts in the feedwater. Successive passes using additional membranes can be added along with other design optimizations to achieve permeate with the TDS required for a target water use. Potable water requirements can readily be met with 200-500 mg/L TDS water, which can be achieved from seawater with a single RO pass.

[c] Cooling water is not factored into these recovery calculations. The MVC process does not require cooling water and typically operates at 23 to 41 percent recovery with seawater desalination, but recoveries can reach as high as 95 percent for industrial concentration/ZLD applications. "Apparent recovery" of thermal desalination that uses intake water used as cooling water can be 10 to 20 percent. However, cooling water volumes can be substantially reduced by employing other cooling mechanisms, such as cooling towers.

SOURCES: Wangnick (2002), German Aerospace Center (2007), GWI (2006a), USBR (2003); Spiegler and El-Sayed (1994); Thomas Pankratz, GWI, personal communication, 2008.

TABLE 4-3 Comparison of Predominant Brackish Water Desalination Processes

	Brackish water RO	ED/EDR	NF
Operating temperature (°C)	<45	<43	<45
Pretreatment requirement	High	Medium	High
Electrical energy use (kWh/m^3)	0.5-3	~0.5 kWh/m^3 per 1,000 mg/L of ionic species removed	<1
Current, typical single train capacity (m^3/d)	<20,000	<12,000	<20,000
Percent ion removal	99-99.5%	50-95%	50-98% removal of divalent ions; 20-75% removal of monovalent ions
Water recovery	50-90%	50-90%	50-90%

SOURCES: Anne et al. (2001), Wangnick (2002), Kiernan and von Guttberg (2005), Reahl (2006), Sethi et al. (2006b), USBR (2003).

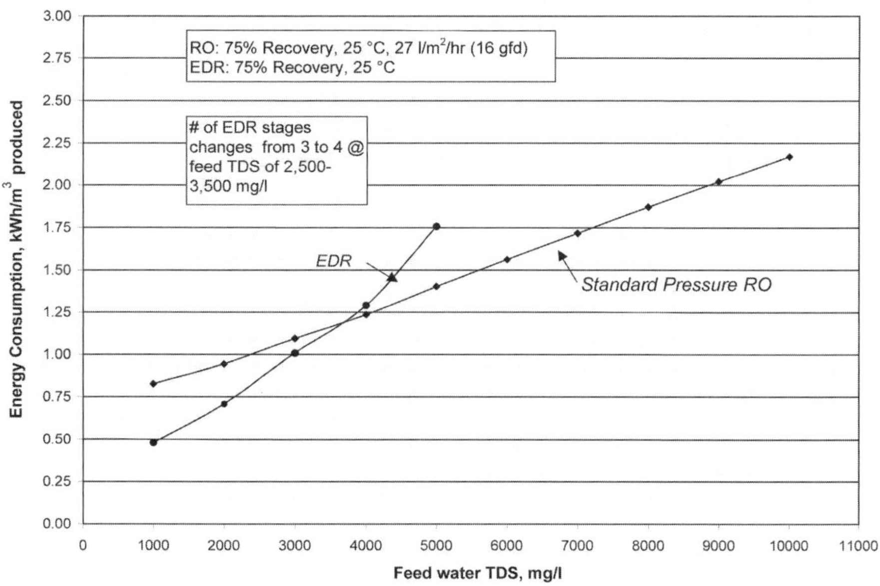

FIGURE 4-10. Comparison of energy consumption by process for the desalination of brackish feedwater across a range of TDS concentrations.
SOURCE: USBR (2003).

polyetheramides, and polyethersulfones. The present most widely used membrane material is a thin-film composite polymer combining a microporous polysulfone support layer with a thin polyamide layer. The membrane, commercialized in 1980, has been hugely successful and shaped the course of membrane technology. RO membranes have matured significantly over the past few decades, with exceptional improvement in RO membrane costs, water flux/permeability, membrane life, and salt rejection capability. Inflation-corrected membrane costs, for example, dropped by a factor of about 4 between 1975 and 1990 and by roughly another 75 percent between 1990 and 2002 (Birkett and Truby, 2007). These improvements in performance, in combination with advances in technologies to recover the unused energy from the concentrate stream, have led to dramatic reductions in energy costs and capital expenses required to desalinate seawater and brackish ground waters.

Although RO technology appears to be maturing, several major challenges remain, including membrane fouling, which leads to increases in energy use and poor resistance to chlorine and other oxidants. Membranes have shifted from the original cellulose acetate membranes to thin-film composite (TFC) membranes. In the past few years, several variations of TFC membranes have been commercialized in an attempt to reduce fouling. Many of these developments have resulted from the addition of polymer to smooth the surface or surface modifications such as addition of different functional groups to change the surface charge. While these improvements reduce fouling, truly fouling-resistant membranes are yet to be realized. Thus, opportunities exist to modify existing or create new membrane formulations or alter surface characteristics to reduce fouling.

Currently, the most direct and effective way to protect against fouling is with effective pretreatment to remove suspended/colloidal matter and dissolved organic matter. As an alternative to fouling-resistant membranes, fouled membranes that could be cleaned easily with low-cost oxidants (e.g., chlorine) would be desirable. However, the state-of-the-art RO membranes for seawater and brackish water desalination cannot tolerate oxidants such as free chlorine, and they require chlorine removal from the feedwater before being processed by the RO modules. Consequently, biofouling can be another challenge that limits the performance of RO membranes.

Another limitation in RO desalination is the relatively low recovery rate in seawater and brackish water desalination (up to about 60 percent and 50-90 percent, respectively), which results in large volumes of concentrate. The maximum recovery is limited by the mechanical pressure limitations of the materials in the membrane element whereas practical recoveries (typically 45 percent for seawater) consider optimization of other parameters such as solubility product limits and energy consumption. Improved membrane materials or configurations may increase the pressure tolerance of the RO modules and/or reduce osmotic pressure of the concentrate due to reduced concentration polarization (i.e., the buildup of dissolved salt near the membrane surface). Several approaches for improvements to overall RO recovery are currently being investigated (Mickley, 2007; Sethi et al., 2008). For example, demonstration testing is under way on a dual RO system with intermediate chemical precipitation to further enhance recovery in brackish water desalination by addressing scaling concerns (Williams et al., 2002). A patented high-efficiency RO technology has also recently been developed that combines a two-phase RO process with chemical pretreatment of primary RO, intermediate ion-exchange treatment of the primary RO concentrate, and high-pH operation of secondary RO to allow operation of the secondary RO at high recoveries (Jun et al., 2004; Mukhopadhyay, 1999). Other examples of high-recovery technologies being developed or tested include dual RO with intermediate biological treatment (Williams and Pirbazzi, 2003), treatment of RO concentrate via EDR with intermediate chemical precipitation (Sethi et al., 2008), and alternative technologies such as such as forward osmosis, membrane distillation, and dewvaporation (see Box 4-5).

Improved module and membrane configuration and system design are also imperative to avoid operation under the phenomenon of "thermodynamic restriction" (Song et al., 2003a). This phenomenon, which is most likely to occur with the use of high-permeability RO membranes, results from significant buildup of salt concentration down the membrane channel (i.e., along the modules in the pressure vessel), such that midway through the channel the osmotic pressure of the concentrate increases to a level equal to the applied hydraulic pressure (Song et al., 2003a, 2003b). Under these conditions, all product water produced by the RO process would permeate out of the membrane before the flow reaches the end of the channel; the rest of the membrane channel would not produce any more water. Song et al. have demonstrated that many RO systems with high-permeability membranes are operated at or near the regime of thermodynamic restriction. Under thermodynamic restriction, increasing the applied hydraulic pressure has little effect on the overall product water flux or water recovery of the RO system. Thermodynamic restriction can be avoided or minimized by proper module configuration and system design, such as membrane modules with higher channel height or the reduction of the number of modules in a pressure vessel.

RO membranes have been dominated by 2.5-, 4-, and 8-inch-diameter spiral-wound thin film composite configurations with a standard length of 40 inches for many years. RO plants that produce between 250 and 330,000 m^3/day currently utilize the 8-inch-diameter membrane,

BOX 4-5
Research on Alternative Desalination Technologies to Improve Energy Efficiency

Several alternative approaches have been proposed to reduce the energy requirements of desalination (see Miller and Mayer, 2005). Some approaches, such as forward osmosis, membrane distillation, and dewvaporation, present opportunities to improve the use of low-grade or waste heat, whereas other approaches, such as freeze desalination and capacitative deionization, offer the potential to reduce overall energy use.

Forward Osmosis

Forward osmosis is a membrane-based separation process that uses osmotic pressure difference between a concentrated "draw" solution and a feed stream to drive water flux across a semipermeable membrane. Given sufficient difference in osmotic pressure, the magnitude of water flux and degree of salt rejection can be competitive with RO (McCutcheon et al., 2005). The primary challenge is in the selection of a draw solute so that its presence in the product water is desirable, or so that it may be easily and economically removed. For example, if a combination of NH_3 and CO_2 gases is used as the draw solution, the energy requirements of the forward osmosis process are small quantities of electrical power (<0.25 kWh/m^3) combined with low-quality heat (<50°C), which could be provided as a reject stream from industrial or power production processes (McGinnis and Elimelech, 2007).

Dewvaporation

In dewvaporation technology, a stream of air is humidified by a falling film of saline water along one side of a heat transfer surface. The air is partially heated by an external source (e.g., solar, waste heat). The heated air then is swept along the condensing side of heat transfer films, where the vapor condenses to a liquid, which is collected as product water. The condensation releases heat through the heat transfer surface to the evaporation side. A small prototype was built and operated, demonstrating the efficacy of the approach (Hamieh et al., 2001). Potential difficulties in the use of this process are the large heat transfer areas required, the impact of ambient weather, and the need for a low-temperature sink to permit condensation. Potential benefits include the efficient use of low-grade heat or solar energy, small footprint, and low capital costs compared to conventional thermal desalination methods.

Membrane Distillation

In membrane distillation, saltwater is warmed to enhance vapor production, and the vapor is exposed to a membrane that can pass water vapor but not liquid water. There are several different types of membrane distillation; the main four types are direct contact, air gap/sweeping gas, osmotic, and vacuum (Banat and Simandl, 1998; Celere and Gostoli, 2002; El-Bourawi et al, 2006; Srisurichan et

continued

al., 2006; Xu et al., 2006). Rejection of feedstream solutes is high and can be comparable to that of other distillation techniques. The possible advantages of the use of membrane distillation are that it has a small footprint relative to other thermal desalination technologies, lower capital costs, and the ability to use low-grade heat sources. Possible disadvantages include difficulty in maintaining the hydrophobicity of the membrane over lcng periods due to fouling and membrane degradation, the large enthalpy of vaporization required for the phase change of water transported across the membrane, and poor rejection of volatile feed-stream contaminants (Peng et al., 2005; USBR, 2004).

Freeze Desalination

The basis of freeze desalination technologies is to change the phase of water from liquid to solid. As ice crystals form, they exclude salt from their structure, enabling the possibility of washing the salt from the crystals. This approach seeks to take advantage of the relatively low enthalpy of phase change—the freezing of water at atmospheric conditions (334 kJ/kg)—whereas evaporation would require 2,326 kJ/kg. The cooling required for the process must be supplied from a means of refrigeration, either mechanical or thermal (absorption cooling). Once crystallization of the water has occurred, the ice crystals need to be separated from the saline solution and washed to ensure final water quality. Potential benefits of the technology include improved energy efficiency compared to distillation processes because ambient seawater is always closer to its freezing point than its boiling point. Potential difficulties include effective separation and washing of water crystals without prematurely melting them and redissolving the salt, maintenance of relatively complex system components, and achieving efficient operation in light of refrigeration requirements. While distillation processes can be cascaded in multiple stages or effects to reuse the latent heat of evaporation (reducing the 2,326 kJ/kg evaporation energy to less than 20 kJ/kg), it is challenging to make a similar freezing configuration.

Capacitive Deionization

Capacitive deionization is an electrosorption process whereby ions are removed from water using an electric field gradient as the driving force. The saline feed flows through electrodes comprised of materials such as carbon-based aerogels. These aerogels have very high surface area (400-1,000 m^2/g) and low electrical resistivity. The cations are attracted to the anodic electrode, while the anions are attracted to the cathodic electrode. A direct current is imparted, with a potential difference of 1-2 volts. Ions are held at the surface of the electrode in the electric double layer. Researchers at Lawrence Livermore National laboratory have determined that this technology can desalinate brackish water (2,000 ppm feed to 186 ppm permeate) using 0.14 kWh/m^3, assuming 70 percent recovery of the stored electrical energy (Farmer et al., 1996).

with tens of thousands of membrane elements required for large desalination plants. By 2004, manufacturers began offering 16- and 18-inch-diameter elements in 40- and 60-inch lengths. Fewer, larger membrane elements may reduce overall capital costs through economies of scale and the need for fewer components (e.g., piping, connections) while reducing the operations and maintenance requirements. The large membrane surface area provided by large-diameter elements would also enable the reduction of the channel length (or the number of membrane elements in the pressure vessel), thereby balancing the permeate through the membrane elements and eliminating operation under thermodynamic restriction where downstream membrane elements do not produce any water flux.

Nanofiltration

Similar to the RO process, the NF process also uses semipermeable membranes and a driving force of hydraulic pressure, in the range of about 50–250 psi. NF membranes are capable of rejecting divalent ions (such as hardness) and larger contaminants very well, while providing lower retention of monovalent ions (see Figure 4-9). NF can also remove synthetic organic chemicals and disinfection by-product precursors. Thus, NF is primarily used for softening and removal of organics. The NF process achieves removal via a combination of both the classic solution/diffusion mechanism as well as steric (size) and charge exclusions (e.g., Childress and Elimelech, 2000; Timmer, 2001). Pilot testing of a two-pass NF system for seawater desalination is under way at the Long Beach Water Department. The first pass removes greater than 90 percent of the salinity, and the second pass removes greater than 93 percent, resulting in a total salt reduction of about 99.5 percent (Tseng et al., 2003). The presence of two passes of NF provides greater flexibility than conventional membrane processes. For example, the second pass can be operated at a higher pH by addition of a base, which allows very high rejection of boron (see also Chapter 5). The overall recovery from the process is about 30 to 45 percent for seawater desalination, which is at the low end of the range observed with conventional RO desalination.

For NF, the typical energy usage is lower than that for RO, depending on the feedwater characteristics and the product water quality objectives. Similar to RO, energy recovery is possible using typical energy recovery devices. As with RO, fouling is a major challenge for efficient operation of NF plants, and pretreatment of feedwaters is needed (USBR, 2003).

Developments in Energy Efficiency for RO and NF

The reduction in energy use for RO in the past 20 years has been remarkable (see Figure 4-11) and has had a significant and direct effect on operating costs. Energy use of as low as 1.6 kWh/m^3 is achievable using controlled, favorable conditions and commercially available state-of-the-art equipment, including energy recovery devices, feed pumps, and low-pressure membranes (Seacord, 2006a). Specific energy values would be larger using real-world conditions.

Energy Recovery Devices. A key reason behind such improvements in the energy efficiency of seawater RO systems has been the development of highly efficient energy recovery devices that capture the energy resident in the concentrate stream of the RO process. Due to low net recoveries of the highly pressurized feedwater, typically 40 to 60 percent of the applied energy in the process can be lost if the concentrate is discharged to atmosphere without any attempt to recover that energy. In general, energy recovery devices can recover from 75 to 96 percent of the input energy in the concentrate stream of a seawater RO plant (Sallangos and Kantilaftis, 2003).

Existing energy recovery systems can be divided into two categories. The first are devices that transfer the concentrate pressure directly to the feedstream (e.g., pressure exchanger, work exchanger), which have energy recovery efficiencies of about 95 percent. The second category includes devices that transfer concentrate pressure to mechanical power, which is then converted back to feed pressure (e.g., Pelton turbine, Francis turbine, reverse-running pumps). The overall efficiency of energy

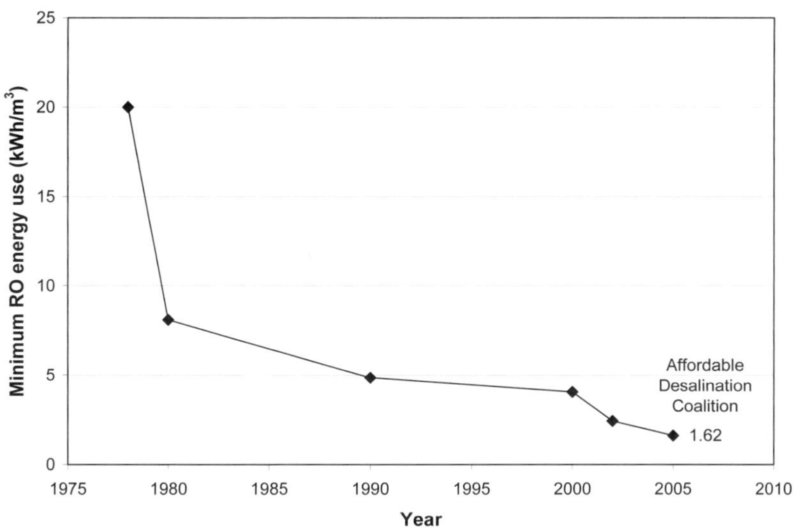

FIGURE 4-11. Seawater reverse osmosis energy use trend.
SOURCES: Data from McHarg and Truby (2004).

recovery here is about 74 percent (assuming a Pelton turbine efficiency of around 87 percent coupled with a pump efficiency of 85 percent). Specific efficiency values for several energy recovery devices are shown in Table 4-4.

Although the pressure exchanger offers higher efficiency than the indirect device group, the choice of energy recovery device for a specific plant design depends on a number of factors. For example, if energy is a critical issue to overall operating costs, the higher-efficiency pressure-work exchangers often will be the device of choice. However, if the project costs are dominated by the capital expenditures, today's pressure-work exchangers have a disadvantage due to their higher equipment costs and current component size limitations (that is, multiple units may be needed at large scales).

Practical Limits to Energy Efficiency of RO. While desalination requires an intrinsic minimum *available* energy as described in Box 4-3, there are practical limits to approaching the minimum required energy of approximately 0.7 kWh/m^3 for infinitesimally small recovery and 0.9 kWh/m^3 for 40 percent recovery. The theoretical minimum energy described in Box 4-3 is pertinent to an ideal reversible process carried out extremely slowly with no energy loses (i.e., applied pressure is only infinitesimally higher than the feed osmotic pressure). In actual desalination processes, energy is lost because of the inherent irreversibility of the processes and a number of practical limitations.

In practice, today's best available RO membranes are operated at pressures significantly above the osmotic pressure to produce practical product water fluxes through the membranes and thereby minimize the net capital expense of the desalination plant. This applied pressure requirement to overcome the osmotic pressure limitations in practice is further elevated to overcome the locally elevated osmotic pressures near the surface of the RO membrane that are caused by high concentrations of salts in the boundary layer near the membrane surface (AWWARF et

TABLE 4-4 Typical Efficiency of Energy Recovery Devices

Energy Recovery System	Efficiency (percent)
Francis turbine	76
Pelton turbine	87
Turbo charger	85
Work exchanger	~96
Pressure exchanger	~96

SOURCES: Geisler et al. (2001), Sallangos (2004), and Lieberman (2003).

al., 1996; Brian, 1966; Sherwood et al., 1964). Further adding to this effect is the increasing bulk salt concentration that results when the feedwater is progressively concentrated as it flows through the membrane pressure vessels (Song et al., 2003). Improvements in the permeability and salt rejection of properties of membranes (see Box 4-6) will also reduce the pressure required to produce practical water fluxes.

Through a simple mass and energy balance calculation (see Appen-

BOX 4-6
Research to Improve Membrane Fouling Resistance, Flux, and Selectivity

Current research efforts, including those in the rapidly growing field of nanotechnology, have the potential to advance technologies for water and wastewater treatment as well as desalination. Examples of efforts involving membrane modification, nanostructures, and nanomaterials for manufacturing new desalination membranes with improved water flux, permeability, fouling resistance, and selectivity are presented.

Membrane Modification to Improve Fouling Resistance

Organic and biological fouling are the result of interactions between solutes and the membrane surface. Thus, surface characteristics of membranes, such as hydrophilicity, surface charge, and roughness, will affect the rate and extent of fouling. Modification of commercially available membranes to alter surface characteristics to reduce fouling while maintaining or improving flux and selectivity is an established research area that shows promising results for RO and NF membranes (Abitoye et al., 2005; Belfer et al., 1998; Gilron et al., 2001). Although many types of modification methods exist, graft polymerization is the method most commonly utilized in RO and NF membranes. If a prudent choice of a monomer is utilized, graft polymerization can increase membrane hydrophilicity and, thus, resistance to fouling with little sacrifice in the flux or selectivity of the membrane.

Carbon Nanotube-Based Desalination Membranes

Theoretical studies and molecular dynamics simulations suggest that hydrophobic channels, like carbon nanotubes, can have considerable water occupancy and that the flow of water in carbon nanotubes is frictionless, limited only by the barriers at the nanotube channel's inlet and outlet. The observed flow rates in the molecular dynamics simulations were quite high, comparable to water flows observed in biological water channels (aquaporins) (Bolhuis and Chandler, 2000; Hummer et al., 2001; Kalra et al., 2003). Inspired by these studies, recent experimental investigations demonstrated high water flows through carbon nanotubes, with values exceeding those calculated from continuum hydrodynamic models by more than three orders of magnitude (Hinds et al., 2004; Holt et al., 2006).

continued

While these experimental observations are promising, no studies have been carried out so far that demonstrate rejection of salt by such nanotube membranes. It is also not clear at this time how such membranes will perform with seawater and brackish waters, where fouling can be an important factor. Finally, even if such nanotube membranes demonstrate desalination performance, scaling to membrane modules and cost of production will remain major obstacles.

Nanocomposite Membranes

Recent efforts aimed at improving the permeability and selectivity of dense polymeric membranes for gas separation demonstrated that dispersion of fumed silica nanoparticles in glassy amorphous polymers can enhance both the permeability and the selectivity of such membranes (Merkel et al., 2002). Mahajan et al. (2002) showed that composite materials comprised of molecular sieve domains (zeolites) embedded in polymer matrices can enhance membrane permeability and selectivity. While the materials and synthesis protocol of gas separation and RO membranes are different (Mulder, 2004), it is possible that nanocomposite RO membranes formed by dispersion of nanoparticles or molecular sieves in polymers would yield enhanced membrane performance. Recently, Jeong et al. (2007) reported the formation of novel thin-film nanocomposite RO membranes incorporating zeolite nanoparticles dispersed within the thin polyamide active layer. The results suggest that the nanoparticles in the active layer can play a role in water permeation and salt rejection. The reported performance data, however, do not exceed the performance of current commercial RO membranes. Further research on refining the synthesis method of thin-film nanocomposite membranes may result in membranes with enhanced water flux.

Biomimetic Membranes

Aquaporins (water channels) and ion channels of biological cells are attracting great interest for their potential to overcome the limitations of polymeric dense membranes and increase water flux and selectivity (Miller and Mayer, 2005), even though there is no published work at this time on biomimetic membranes for engineered applications. Cell membranes of animals and plants are highly selective barriers that regulate the transport of water, ions, and uncharged solutes into the cell by means of specialized protein channels—*aquaporins* for transport of water and *ion channels* for regulating the transport of ions (Borgina et al., 1999). The most unique aspect of the water and ion channels is their very high selectivity. Aquaporins allow only water molecule transport through the protein channel at flow rates several orders of magnitude larger than expected for a channel of only a few angstroms in diameter. Similarly, ion channels are highly selective structures in membrane cells, allowing for the selective transport of certain ions. No synthetic analogues have been developed with water permeability and ion selectivity as high as those found in aquaporins and ion channels, although Walz et al. (1994) incorporated aquaporin proteins into a lipid bilayer membrane that exhibited extremely high water permeability. Development of synthetic analogues for aquaporins or incorporation of aquaporins within a membrane matrix may lead to a major advancement in current desalination membrane technology.

dix B), however, the committee concludes that the practical upper limit of energy savings that can be realized through advances in RO membranes is approximately 15 percent. This estimate was made by assuming a system operating at 40 percent recovery with a 95 percent energy recovery device and a new advanced seawater RO membrane with twice the permeability of today's best available membranes while not sacrificing salt rejection characteristics—a huge advancement above today's best available technology. This simple analysis implies that the RO process is approaching a state of diminishing returns as it relates to energy usage. Although these improvements would still provide a cost savings to the desalination process, an improvement in energy savings beyond 15 percent appears to be a significant challenge. Improvements in module design that enable operation at higher fluxes appear to have the greatest potential for reducing the overall operating costs of desalination because the capital costs and energy costs per cubic meter of permeate produced would simultaneously be reduced (see Box 4-6). Alternatively, a breakthrough in an alternate technology to RO may allow even greater energy savings (see Box 4-5).

Electrodialysis and Electrodialysis Reversal

The ED and EDR processes use ion-selective membranes and an electrical potential driving force to separate ionic species from water. Ionic species are driven through cation- and anion-specific membranes in response to the electrical potential gradient while the ion-depleted water passes between the membranes. The EDR process is similar to the ED process, except that it also uses periodic reversal of polarity to effectively reduce and minimize scaling and fouling, thus allowing the system to operate at comparatively higher recoveries. As of 2005, ED represented over 3 percent of the worldwide online desalination capacity and nearly 8 percent of online capacity in the United States (GWI, 2006b).

EDR and ED processes are typically used to desalt brackish water—not seawater—because the cost of these processes increases significantly with higher salinity or TDS (Figure 4-10). In general, municipal applications for ED/EDR have been noted for brackish waters with TDS up to 7,500 mg/L (Mickley et al., 2006), although ED/EDR is typically cost-competitive with RO for TDS up to about 3,000 mg/L (Mallevialle et al, 1996). As with other membrane processes, ED membranes are subject to fouling and some pretreatment of the feedwater is necessary. Typically, prefiltration is required to remove suspended solids and CO_2 is removed to improve energy efficiency (Kiernan and von Guttberg, 2005; Reahl, 2006; Weber, 1972).

Even though the ED/EDR process is more energy intensive than RO with source water above approximately 3,500 mg/L TDS (Figure 4-10), the process still maintains an important niche in desalination technologies. Unlike RO and thermal desalination processes, ED is only capable of removing ionic components from solution. As a result of this phenomenon, fouling by uncharged species like silica is less severe as compared to the RO process. Additionally, current ED/EDR membranes are resistant to chlorine, making them more robust for processing feedwaters with higher levels of organic matter that would typically foul RO membranes (e.g., water reuse applications). These features are important factors that increase the practical application of ED/EDR over RO for such challenging applications. Because of the robust nature of EDR, it is also being applied in hybrid applications as a concentrate reduction method for RO processes (Kiernan and von Guttberg, 2005; Reahl, 2006).

Thermal Desalination Processes

Thermal distillation was the earliest method used to desalinate seawater on a commercial basis, and thermal processes have been and continue to be a logical regional choice for desalination in the Middle East for several reasons. First, the seas in the region are very saline, hot, and periodically have high concentrations of organics, which are challenging conditions for RO desalination technology. Second, RO plants are only now approaching the large production capacities required in these regions. Third, dual-purpose cogeneration facilities were constructed that integrated the thermal desalination process with available steam from power generation, improving the overall thermodynamic efficiency by 10-15 percent (Hamed et al., 2002; Hanafi, 2002). For these reasons combined with the locally low imputed cost of energy, thermal processes continue to dominate the Middle East. In other parts of the world, where integration of power and water generation is limited and where oil or other fossil fuels must be purchased at market prices, thermal processes are relatively expensive (GWI, 2006a).

In the United States, thermal processes are primarily used as a reliable means to produce high-quality product water ($\leq$ 25 ppm TDS) for industrial applications, because distillation processes are very successful at separating their target—dissolved salts—from the bulk feedwater. Distillers almost completely reject dissolved species, such as boron, which can be problematic for RO. Distillers, however, are sensitive to volatile contaminants that may evaporate from the feedwater and carry over into the distilled water, where they may or may not condense.

Three major thermal processes have been commercialized: MSF distillation, MED, and MVC, and each is a mature and robust technology (see Box 4-7 and Table 4-2). MSF and MED processes demand both thermal energy (typically steam) and electrical energy. Thermal processes are configured to use and reuse the energy required to evaporate water, known as the latent heat of evaporation (about 2,326 kJ/kg of water or 644 kWh/m^3 at normal atmospheric conditions). How efficiently the latent energy is reused is a function of project-specific economics, considering capital and operating costs.

BOX 4-7
Overview of Thermal Desalination Processes

Three primary thermal desalination processes have been commercially developed:

- **Multistage flash** (MSF) distillation, a forced circulation process, is by far the most robust of all desalination technologies and is capable of very large production capacities per unit. Globally, MSF is among the most commonly employed desalination technologies. MSF uses a series of chambers, or stages, each with successively lower temperature and pressure, to rapidly vaporize (or "flash") water from the bulk liquid. The vapor is then condensed by tubes of the inflowing feedwater, thereby recovering energy from the heat of condensation (Figure 4-12). The number of stages used in the MSF process is directly related to how efficiently the system will use and reuse the heat with which it is provided.

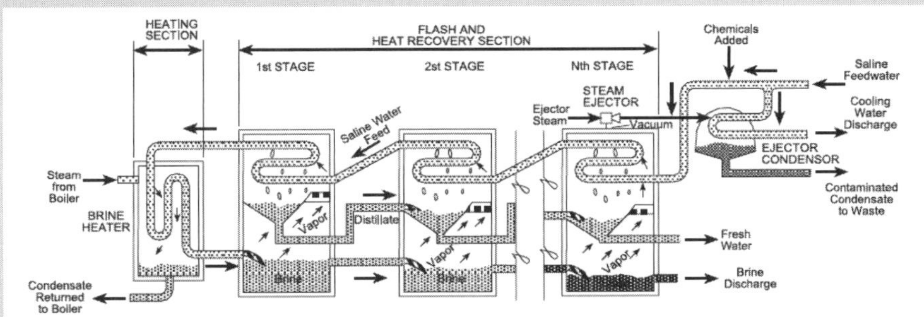

FIGURE 4-12. Multistage flash evaporation.
SOURCE: Buros et al (1980); Buros (2000).

- **Multiple effect distillation** (MED) is a thin-film evaporation approach, where the vapor produced by one chamber (or "effect") subsequently condenses in the next chamber, which exists at a lower temperature and pressure, providing additional heat for vaporization (Figure 4-13). MED technology is being used with increasing frequency when thermal evaporation is preferred or required, due to its reduced pumping requirements and thus its lower power use compared to MSF. MED plants were initially limited in size but MED technology is planned for an 800,000 m^3/day desalination plant in Jubail, Saudi Arabia.

continued

Since the early 1990s, MED has been the process of choice for industrial low-grade-heat-driven desalination. The largest MED plants incorporate thermal vapor compression (TVC), where the pressure of the steam is used (in addition to the heat) to improve the efficiency of the process.

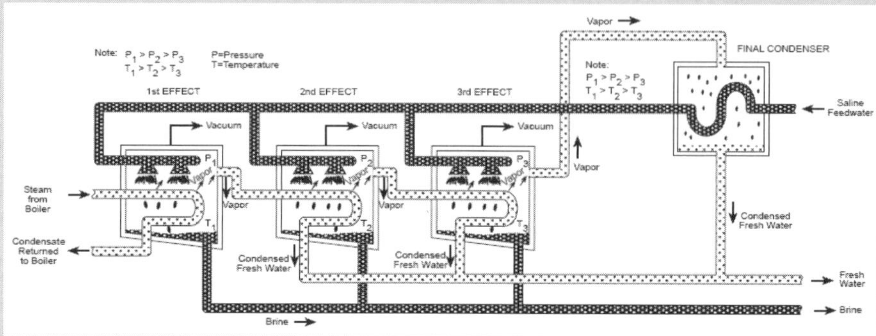

FIGURE 4-13. Multiple effect distillation process.
SOURCE: Buros et al (1980); Buros (2000).

- **Vapor compression** (VC) is an evaporative process where vapor from the evaporator is mechanically compressed and its heat used for subsequent evaporation of feedwater (Figure 4-14). VC units tend to be small plants of less than 2,839 m^3/day that are used where cooling water and low-cost steam are not readily available. VC systems can operate at very high salt concentrations and the VC process is at the heart of many industrial zero liquid discharge systems (Pankratz and Tonner, 2003).

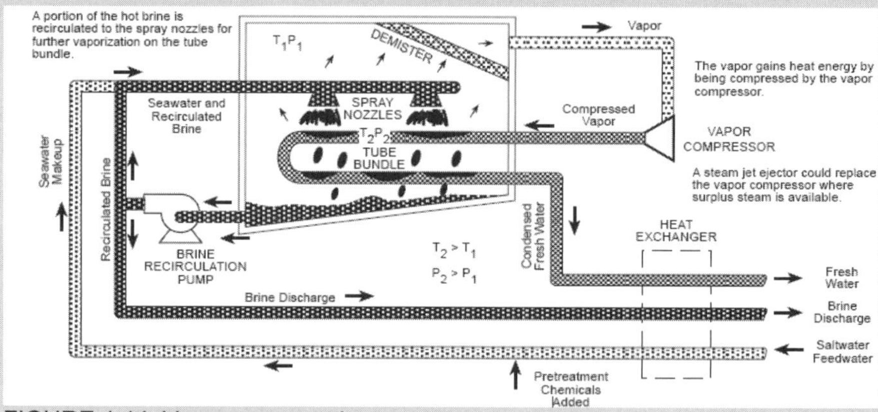

FIGURE 4-14. Vapor compression process.
SOURCE: Buros et al (1980); Buros (2000).

Other nonhybrid thermal desalination approaches, including solar stills and freezing, have been developed for desalination, although they have not been commercialized to date (Buros, 2000). In brief, a solar still uses the sun's energy to evaporate water from a shallow basin, which then condenses along a sloping glass roof. Freezing technologies are described in Box 4-6.

The combined energy requirements of thermal technologies are greater than that of membrane processes, but it is not simple to compare the total energy use of these diverse processes, because MSF and MED are capable of using low-grade and/or waste heat, which can significantly improve the economics of thermal desalination (see Box 4-8). Utilities in the United States have generally overlooked opportunities to couple thermal processes with sources of waste heat to produce desalinated water more economically. In the Middle East, the largest of the MSF and MED plants are built along with power plants and use the low-temperature steam exhausted from the power plant steam turbines. This "cogeneration" approach combines water production with the generation of electric power using the same fuel and offers a method to improve the energy efficiency of desalination plants while sharing intake and outfall structures. Large MSF distillers are commonplace in the Middle East largely because of cogeneration. In another example, many of the largest modern cruise ships select the thermal MED desalination process be cause MED requires 20 to 33 percent of the electrical energy of RO and because the heat energy it requires can be obtained from the ships' propulsion engines. MSF and, increasingly, MED units are also used in industry to make water for liquid natural gas and methanol plants. These industrial processes have a relatively small demand for freshwater relative to the massive quantities of waste heat generated by the petrochemical process and can be designed to be quite inefficient. When the residual heat energy has little or no value, there is no economic justification to invest in more efficient designs.

Scale deposition in thermal desalination units is a concern but is generally mitigated by control of the operating temperatures and concentrations and use of polymer-scale inhibitors. The potential for mineral-scale deposition in a thermal desalination plant is an economic optimization issue, not a limitation of the process. Thermal technologies, including variations of MSF's forced circulation configuration, can work with supersaturated salt solutions and are used in brine concentrators for minimizing the volume of desalination concentrate. However, operating at extremely high recoveries is not usually economical for desalination applications due to the boiling point elevation caused by the salt. In fact, economic considerations affected by boiling point elevation normally limit water recovery of thermal seawater desalination plant designs to about 35 to 50 percent, not considering cooling water.

Although thermal desalination technologies are mature technologies, opportunities remain for additional cost savings. Thermal technologies are not optimized to the highest efficiencies due to current practical constraints in materials and design and considerations of the source, condition, and value of the thermal energy being utilized. All thermal proc-

BOX 4-8
Low-Grade and Waste Heat for Desalination

Low-grade heat and waste heat are two terms that are often used synonymously but, depending upon the application, they may have completely different meanings. The term *low-grade heat* is often used to describe heat energy that is available at relatively low (near-ambient) temperatures that is of minimal value for industrial or commercial processes. In contrast, waste heat, which may or may not be low-grade heat, contains energy that is released to the environment without being used. Both have potential value for desalination.

Most of the largest desalination facilities in the world are dual-purpose facilities that produce both freshwater and electricity. In all of these facilities at least some of the electricity is generated by high-pressure steam when it is expanded through turbines. In the case of backpressure turbines, when the steam leaves the turbine, it can no longer produce electricity even though it is still slightly above atmospheric pressure. The waste energy from this exhaust steam is ideal for use by thermal desalination processes. In contrast, condensing turbines have a cool exhaust steam under vacuum conditions. Therefore, when condensing turbines are used in combination with thermal desalination, some low-pressure steam is extracted for use in the desalination process. Extracted low-grade steam could, in theory, be used to generate more electricity, but practical circumstances (e.g., electricity demand, limited operating flexibility) influence whether this low-grade energy would, in fact, be used this way. Thus, low-grade heat might also be wasted under specific circumstances. Large slow-speed diesel generators, such as those used to power large ships, also represent a source of low-grade heat that is often wasted. The cooling water can easily be used to heat both MED and MSF processes without affecting the efficiency of the power generation. Exhaust-gas boilers can also be added to capture otherwise wasted energy for use for desalination or to generate additional electricity.

There are other potential sources for waste heat that are simpler to identify as waste, such as industrial stack emissions or cooling circuit heat that is rejected to rivers, lakes, or the air via heat exchangers or cooling towers. Contrary to common belief, these heat plumes may contain useful energy, even though this energy may not be economical for use in the existing industrial processes.

There are economic costs associated with the use of waste or low-grade heat, such as the cost of installing and operating the heat recovery system. The act of recovering the heat may also affect the efficiency of the main process. When a previously wasted energy stream is used, it may then be valued as a potential revenue stream by its owner. When these costs are considered, the energy is not free, but in many cases energy costs can be reduced to a small fraction of the total process cost of desalination.

esses are affected by the cost of heat transfer surfaces—which are primarily copper or titanium alloys—and could benefit from development of new material options. Also, the methods of distributing feedwater over the heat-transfer surface of thin-film processes (e.g., MED, MED-TVC, VC) are proprietary and could benefit from development. There may be additional opportunities for improved efficiencies in new designs of thermocompressors for MED-TVC systems.

There are also needs for additional research and development into improved configurations and applications to utilize low-grade and/or waste heat and into entirely new processes that optimize the use of low-grade heat (see Box 4-5). For example, there has been a recent review (Shih and Shih, 2007) of an industrial application that would utilize low-grade energy at sulfuric acid plants. Heat is produced when sulfur is burned and when concentrated acid is diluted. Thermal desalination plants incorporated into this process could therefore produce the water used to dilute the acid which in turn produces the heat required for the thermal desalination process. The location of low-grade and/or waste heat resources near saline water sources and large consumers of water, including industry, has not been investigated, and research on opportunities to utilize low-grade and/or waste heat could yield economical applications of existing thermal desalination technology in the United States.

Ion Exchange

Ion exchange is mainly used for water softening and demineralization, and applications of ion exchange at the municipal level are limited. In an ion-exchange process, water can be desalted by first passing it through a column of cation exchanger beads in the hydrogen (H^+) form. Hydrogen ions replace the cations in the solution, which become bound to the exchanger. The water is then passed through a column of anion-exchange beads in the hydroxyl (OH^-) form where the anions replace the hydroxyl ions, which in turn react with the hydrogen ions in the water. This process can produce almost completely deionized water. When exhausted, the exchangers can be regenerated—the cation exchanger with acid and the anion exchanger with base. The problem is that removal of 1 pound of salt takes about 1.5 pounds of acid and 1.5 pounds of base to regenerate the exchangers (Xu, 2005).

This process makes economic sense compared to other desalination processes only where there is a small amount of salt to be removed from the water. Therefore, the major application of ion exchange has been in the field of production of ultrapure water. Ion exchange is often used as a "polishing" step following another desalting process. Thus, ion exchangers alone cannot economically be used for desalination of seawater or brackish water.

Hybrid Configurations

Hybrid desalination configurations include combinations of processes designed to improve process efficiency or reduce energy costs. Hybrid thermal-membrane facilities incorporate both thermal and membrane desalting processes that are typically co-located with a power plant to improve overall process economics. One hybrid approach blends the product water from parallel RO and thermal desalination processes, which enables the RO membranes to operate with higher permeate TDS (Ludwig, 2004) and which can reduce the replacement costs of RO membranes by up to 40 percent (Hamed, 2005). Hybrid thermal-membrane facilities can also optimize water production and energy costs under seasonal variations in power loads, because operation can be switched from electrically driven RO to thermally driven distillation. In periods of high power demand there will also be associated abundant steam generation such that thermal desalination operations can be maximized; conversely, when there is low power demand (and reduced quantities of available low-grade or waste heat), water production by RO is likely to become more economical (Ludwig, 2004). However, to utilize this operational flexibility and realize the cost benefits, the total installed capacity must be larger than the nominal demand for water. Fujairah in the United Arab Emirates is one facility of this type, with a total water production of 454,000 m^3/day (120 MGD). Two-thirds of the production capacity is provided by MSF units, with the remaining capacity provided by seawater and second-pass brackish-water RO units. The facility also has the capacity to use warm MSF cooling water as part of the feedwater to increase the permeability of the RO process during winter months.

Hybrid desalination facilities may also integrate multiple processes in series to increase the separation or concentration capabilities of the facility. These series hybrids are typically smaller in capacity. For example, zero liquid discharge (ZLD) systems (i.e., facilities with no offsite liquid-waste discharges) often concentrate the desalination waste stream by separating the process into logical steps and optimizing the entire system, using RO systems followed by distillation concentrators and crystallizers. Another hybrid example is the combination of ED and RO proposed by Davis (2006). The process uses ED to reduce the salinity of the reject stream from the RO so that the salt-depleted reject stream can be recycled to the RO to increase recovery. Hybrid configurations in series can also be used to create ultrapure water required by some industrial processes. The multitude of possible combinations of desalination processes in hybrid configurations is limited only by ingenuity and the identification of economically viable applications.

Assessment of Desalination Process Technology

The major desalination technologies currently in use are generally efficient and reliable, but the cost and energy requirements remain high. Ongoing research efforts are motivated by the need to reduce cost or to overcome operational limits of a process, such as reducing membrane fouling or increasing energy efficiency. Although existing desalination technologies will continue to see incremental improvements, the current technologies are relatively mature, and the practical limits of further energy savings through advances in RO membranes is approximately 15 percent. Thus, alternatives to the major desalination technologies continue to be investigated to enhance or replace existing desalination processes or fill niche applications where mainstream technologies are inapplicable (see Boxes 4-5 and 4-6). As discussed earlier, no desalination process can overcome the thermodynamic limit of desalination in terms of energy use (see Box 4-3). Nevertheless, research on alternative desalination technologies is under way in hopes of more closely approaching the thermodynamic energy limit or finding ways to power the desalination process with less-expensive energy sources, such as low-grade heat.

POST-TREATMENT

Desalinated water, produced directly from either thermal or membrane processes, is significantly stripped of dissolved solids, which results in a water quality that has low hardness and alkalinity. Consequently, without proper post-treatment, this water would be corrosive to pipeline materials, including metals and concrete, and may introduce metals into drinking water and reduce the lifetime of water-system infrastructure. Current technology enables desalinated water to be made noncorrosive by adding chemicals such as calcium hydroxide (slaked lime) to increase the hardness and alkalinity and sodium hydroxide (caustic soda) to adjust the pH. Carbon dioxide is commonly used to normalize the pH. Post-treatment of water is a mature science. The water chemistry issues are generally well understood and methods of altering chemical conditions are feasible and generally available, although the exact process used will depend greatly on the particular chemistry of the desalinated water and the existing infrastructure (see also Chapter 7).

CONCENTRATE AND RESIDUALS MANAGEMENT

All desalination processes leave behind a concentrated salt solution that may also contain some pretreatment and process residuals (see Tables 4-1 and 5-1). Concentrate and residuals management involves waste minimization, treatment, beneficial reuse, and disposal, and conventional concentrate management approaches are described in this section. Each approach has its own set of costs, benefits, environmental impacts, and limitations (Sethi et al., 2006a). Further, state regulations may limit the concentrate management practices available at any individual site (Mickley, 2006). Because of the widely varying level of technology involved in concentrate management options and site-specific factors and regulatory considerations that limit available alternatives, the cost of concentrate management can range from a relatively small fraction of the cost of the main desalination system to as high as several times the cost of the desalination system.

The state of the technology, including advantages and disadvantages, for each of the current methods of concentrate management is discussed in this section. A summary of the challenges and limitations in the current state of concentrate management methods is also provided in Table 4-5. The environmental impacts of concentrate management alternatives are discussed separately in Chapter 5.

Surface Water Discharge

Surface water discharge to a receiving body is the most common concentrate management practice in the United States, employed by approximately 41 percent of municipal desalination facilities greater than 95 m^3/day (Mickley, 2006; Figure 4-15) and at all seawater desalination facilities of significant capacity worldwide. Direct surface water discharge of concentrate is a relatively low-energy, low-technology solution to concentrate management. Costs are generally low assuming that the length of the pipeline is reasonable and the concentrate meets the permit requirements without the need for further treatment. However, it has the potential for negative impacts on aquatic organisms (see Chapter 5) and for complex permitting requirements. The salinity of the concentrate is typically higher than that of the ambient water, but good site location, engineering practice, diffuser design, and/or dilution with additional water (or treated wastewater) prior to discharge can likely minimize most potential negative environmental effects.

Multiport discharge diffusers (Figure 4-16) are being employed at some seawater desalination plants to minimize environmental impacts. Studies show that concentrate, being denser than seawater, may sink and impact benthic communities. By employing multiple outlet ports, rather than a single open pipe, the mixing and dilution of the concentrate can be accelerated, lessening potential impacts in sensitive areas (EPA, 1991).

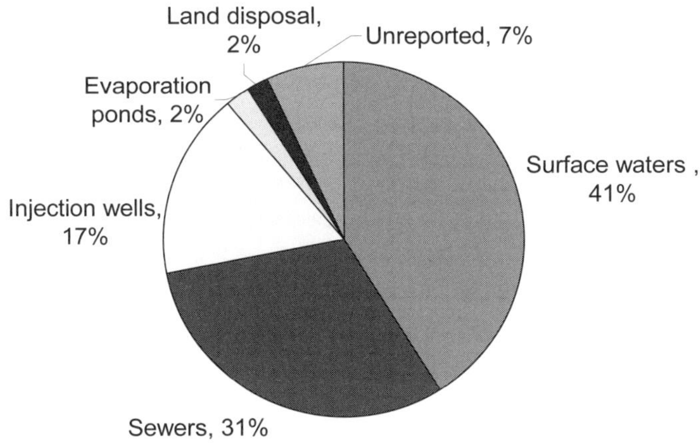

FIGURE 4-15. Identified methods of concentrate management, based on a survey of the 234 municipal desalination plants in the United States with output greater than 95 m^3/day (25,000 gallon per day).
SOURCE: Data from Mickley (2006).

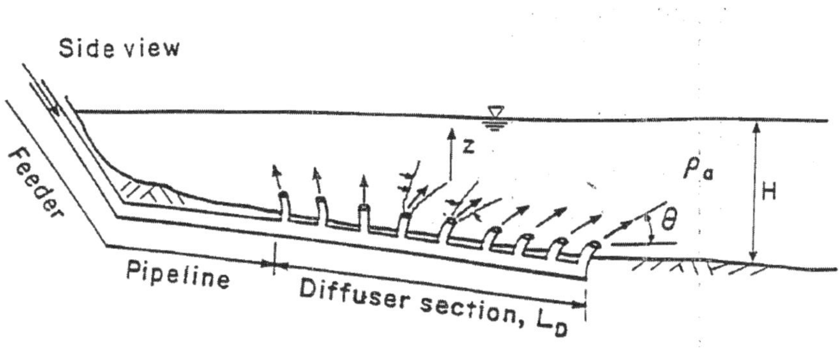

FIGURE 4-16. Multiport diffuser for improved initial mixing of surface water concentrate discharge.
SOURCE: EPA (1991).

Blending and diluting the concentrate with treated wastewater or power plant cooling water is also desirable to reduce environmental impacts but is not an option for all site locations. Shipboard desalination configurations can blend the concentrate with large volumes of ocean water, taken in specifically for that purpose.

Sewer Discharge

Discharge of concentrate into an existing sewage collection system is the second most common concentrate management practice in the United States, employed by approximately 31 percent of surveyed municipal desalting facilities (Mickley, 2006). This method is also relatively low in cost and in energy use but retains the potential for adverse environmental impacts from elevated concentrations of salt or trace elements in the treated effluent (Table 4-5). A permit from the local sewage agency is required to ensure that potential adverse impacts on the wastewater treatment processes, if any, are within acceptable limits. Large-volume discharges are typically not practical or suitable.

Subsurface Discharge

Concentrate discharge via a subsurface discharge structure, such as a deep well, can occur in both inland and coastal areas. Deep-well injection is a mature technology that involves the disposal of concentrate into a deep geological formation, usually inland, that will serve to permanently isolate the concentrate from aquifers that may be used as a drinking water source. Appropriate geology with the presence of a structurally isolating and confining layer between the receiving aquifer and any overlying source of drinking water is required. Suitable formations for injection often contain water with TDS concentrations in excess of 10,000 mg/L. These conditions are determined through site-specific hydrogeologic assessments. Deep-well injection is commonly employed by desalination plants in certain parts of Florida and island applications where receiving aquifers can be found at relatively shallow depths, but it is less common elsewhere in the United States. Deep-well injection is used at 12 percent of municipal desalination facilities in the United States with output greater than 95 m^3/day (Mickley, 2006). It is typically employed for larger desalination plants (e.g., >3,800 m^3/day [>1 MGD]) because the costs for developing deep-injection wells are relatively high and are not largely reduced for smaller flows. For example, the typical capital cost of a 3,000-m-deep well is reported at $8.1 million for a concentrate flow of 3,800 m^3/day, which decreases to only about $5.1 million for a concentrate flow of either 380 or 38 m^3/day (Malmrose et al., 2004). These costs exclude any pretreatment or standby disposal system. While capital costs for well injection are about average of typical inland concentrate management methods, the annual operating costs are relatively low as a percentage of total operating costs (Mickley, 2006). For example, the capital cost of well injection and several tens of miles of delivery pipes was 26 percent of total capital costs for the desalination project, but the operating costs are estimated to be 4 percent of the annual operating cost for the El Paso Water Utilities desalination facilities (Ed Archuleta, El Paso Water Utilities, personal communication, 2006; see also Box 5-2). If permitted by state regulations, existing wells from depleted oil and gas reservoirs could be used for concentrate injection, although their injection capacity would need to be evaluated, and the costs of transporting the concentrate may offset other cost savings.

For seawater desalination, subsurface discharge involves using a beach well or percolation gallery beneath the beach or underneath the seafloor. Because mixing occurs beneath the surface and the discharge plume slowly dissipates into the surf zone, subsurface coastal discharge can be an effective way to minimize environmental impacts, although it requires specific hydrogeological conditions.

Land Application

Land application of brackish desalination concentrate can be used for lawns, parks, golf courses, or crop land; it is not practical for the large volume and highly saline concentrate from seawater desalination and is thus only considered for brackish water applications. Even for these applications, a TDS greater than about 5,000 mg/L in the concentrate can typically preclude spray irrigation (Mickley, 2006); thus, there is typically a need for addition of dilution water. Land application is usually practical and employed only for smaller concentrate flows and, because irrigation demands are seasonal, a second or backup disposal or storage method is also necessary for year-round operation (Malmrose et al., 2004). The key concerns with spray irrigation include the influence of concentrate on the soil and vegetation, potential contamination of groundwater, and runoff to surface water (see Chapter 5). The allowable salinity will depend on the tolerance of target vegetation, percolation rates, and the ability to meet the groundwater quality standards. In general the vegetation used is dependent on the site location; however, typically grasses that are high in water and salinity tolerance are used with concentrate discharge. Research is now under way to develop genetically modified feed crops that would tolerate and take up more salt.

TABLE 4-5. Concentrate Management Challenges and Limits

Method	Capital Costs[a]	O&M Costs[a]	Land Area Required	Permitting Complexity	Applicable for Large Conc. Flows	Potential Environmental Impact
Surface water discharge	L[a]	L[a]	--	H	Yes	M
Sewer discharge	L[a]	L[a]	--	M	No	M
Subsurface discharge (deep well injection)	M-H	M	L	M	Maybe	L
Evaporation Pond	H	L[c]	H	M	No	M
Land Application	M	L	H	M	No	M-H
Thermal evaporation	H	H	L	L[d]	No	L[d]

L = low; M = medium; H = high; dashes indicate not applicable.

[a] Costs are highly site-specific; general trends in relative costs are indicated; cost for surface water or sewer discharge can be higher if the distance from desalination facility to the discharge water body or sewer is large, necessitating long pipelines and/or pumping facilities.

[b] Energy use for surface water or sewer discharge or land application can possibly be higher if the distance from desalination facility to the discharge water body, sewer, or land application site is large, possibly necessitating pumping facilities.

[c] O&M costs for evaporation ponds can possibly be higher if a significant amount of monitoring wells and associated water quality analysis are required.

[d] Permitting complexity and environmental impacts of thermal evaporation can possibly be higher if the feedwater-to-desalination process contains contaminants of concern that could be concentrated to toxic levels in the concentrated slurry or solids that are produced from this concentrate treatment process.

[e] Low (L) pertains to Florida (where deep-well injection is commonly practiced) and moderate (M) pertains to other states in the United States.

[f] Climate can indirectly influence surface water discharge by affecting the quantity of surface water available for dilution.

Evaporation Ponds

Evaporation ponds are a low-technology but high-cost approach to concentrate management, where the concentrate is pumped into a shallow lined pond and allowed to evaporate naturally using solar energy. In evaporating environments, the thermodynamic activity of the concentrate decreases with increasing concentration and will approach the average relative humidity of the air. At that point, effective natural evaporation will cease—that is, an evaporation pond that is not leaking will not

Possible Pre-treatment Needs	Labor Needs and Skill Level (for operation)	Energy Use	Public Perception Concerns	Climate Limitation	Special Geological Requirements
M	L	L[b]	H	Maybe[f]	N
L	L	L[b]	L	N	N
L	L	M	L-M[e]	N	Y
L	L	L	H	Y	Y
L	L	L[b]	H	Y	Y
L	H	H	L	N	N

evaporate to dryness in most environments. Periodic removal and drying of accumulated solids is necessary for long-term physical sustainability of a site, although some evaporation ponds are closed and sealed once the pond is filled by solids. Evaporation ponds, under suitable climatic conditions, enable operation of the desalination plant under ZLD conditions, where no liquid waste leaves the plant boundary. Evaporation ponds can be a viable option in relatively warm, dry climates with high evaporation rates, level terrain, and low land costs. They are typically practical and employed only for smaller concentrate flows and are often coupled with high-recovery desalination processes. This disposal method has high capital costs due primarily to the land acquisition costs to accommodate the large surface areas required and also the costs of impermeable liners, if needed. For example, assuming a relatively high evaporation rate of 0.1 $L/h/m^2$, the typical capital cost of an evaporation pond is reported at $40 million for a concentrate flow of 3,800 m^3/day (1 MGD), reducing to $4 million for a concentrate flow of 380 m^3/day (Malmrose et al., 2004). The costs for a lower evaporation rate would be proportionally higher (e.g., four times as much for a rate of 0.025 $L/h/m^2$). These costs exclude any solids disposal or seepage monitoring.

Thermal Evaporation

A thermal evaporator (also known as a brine concentrator) can reduce desalination concentrate to a slurry of approximately 20 percent solids. Thermal evaporators are generally based on vapor compression technology (See Box 4-7). Most evaporators in operation are of the vertical-tube, falling-film type and employ a calcium sulfate seeded slurry process to prevent scaling (Mickley, 2006). Thermal evaporation uses a large amount of energy—more than 18.5 kWh/m^3 of feedwater (Bond and Veerapaneni, 2007). Thermal evaporators have been used in industrial RO applications and are known to be a viable and reliable technology.

When an evaporator is followed with a crystallizer or spray dryer, the concentrated slurry can be further reduced to solids that are suitable for landfill disposal (a zero liquid discharge (ZLD) approach). Like thermal evaporators, crystallizers are also driven by vapor compression. In a crystallizer, the brine enters the vapor body at an angle and swirls into a vortex. As water evaporates, the salt crystals are separated using a centrifuge or filter. Spray drying is another means of producing dry product from concentrate. In this method the concentrated salt solution is reduced to a fine spray and mixed with a stream of hot gas, which provides the heat for evaporation and carries off the moisture released from the concentrate. The resultant dry salt powder is collected in a bag filter. The crystallizer and spray drying processes are more capital-cost- and energy-intensive than the brine concentrator. The energy requirements of thermal evaporators combined with crystallizers can exceed 32 kWh/m^3 (Pankratz, 2008).

These thermal technologies may be coupled with other membrane-based high-recovery processing technologies, described earlier in the chapter, to reduce the overall energy requirements. In a pilot study of five inland brackish water sources, Bond and Veerapaneni (2007) were able to reduce the total energy use of desalination with ZLD to 0.45-1.9 kWh/m^3 of product water by developing a process train involving two RO passes, intermediate concentrate treatment, and a brine concentrator, followed by an evaporation pond.

In general, thermal evaporation-based processes are characterized with high capital costs and energy requirements. The capital and operating costs of these thermal evaporation methods can sometimes exceed the cost of the desalting facility (see Chapter 6). Additionally, once most or all of the liquid is removed from the wastes, landfilling costs can be significant. The high costs, including high-energy requirements, are a large deterrent to application of this process, particularly for large municipal applications. Thus, at the present time, ZLD concentrate management approaches are typically not considered for municipal drinking water applications. Nevertheless, thermal evaporation processes are being considered for some water supply applications in inland regions where the concentrate flows are small and other methods of concentrate management are not feasible (e.g., the desalination facility at the Deuel Vocational Institution in Tracy, California, incorporates a brine concentrator). For thermal evaporation applications to become more viable, improvements are needed that reduce capital costs and/or energy usage.

Development of beneficial salt reuse options and specific salt separation methods are also important to cost reduction of the overall process (Drioli et al., 2004). Some examples of beneficial reuse of the solid product include extraction of gypsum and sodium chloride by means of selective precipitation. However, the economic viability of beneficial reuse of desalination by-product salts depends on finding local markets to avoid high transportation costs (Jordahl, 2006).

Residuals Management

In addition to the concentrate, there are other waste streams from desalination that need to be managed. The state of the science here is relatively mature. The spent cleaning solutions are either disposed of with the concentrate or separately, usually into a sewer system. In the latter scenario, some pretreatment in the form of neutralization for pH adjustment may be required.

In the pretreatment step, solids are removed. In some cases, these solids can be recombined with the concentrate discharge and disposed to the source water. However, more commonly the solids are separated with clarifiers and sent to a belt press for further dewatering. The resulting sludge is then hauled to a landfill. The use of a microfilter will reduce the volume of sludge to be settled in the clarifier as long as the mass of chemicals required to flocculate solids is reduced over conventional processes.

The membranes and security filter cartridges also constitute a residual when they reach the end of their effective life. These residuals are commonly disposed in landfills. A few companies recover used membranes and clean them for further use in a different application.

CONCLUSIONS AND RECOMMENDATIONS

Although RO and thermal-based processes are relatively mature, opportunities exist to improve the energy efficiency and reduce costs. For membrane-based desalination, the most significant improvements can be realized through improved pretreatment and the creation of low-fouling, high-flux (i.e., high-permeability) membranes that can operate at lower pressures. For thermal-based desalination, costs can be reduced by more effective use of low-grade and/or waste heat. Improvements in concentrate management would have both cost and environmental implications. These and other research findings are summarized in this section.

RO technology is approaching the thermodynamic minimum energy value. The membrane industry has made great strides in reducing energy use for the desalination process with the commercialization of high-efficiency energy recovery devices and improvements in membrane technology. Current energy use is within a factor of 2 of the theoretical minimum value for seawater desalination. Practical and economic constraints are likely to inhibit RO energy use from decreasing more than approximately 15 percent from current values. This level of improvement would still be valuable to reducing cost and energy use, but greater returns on investment than this should not be expected. RO is the standard by which novel technologies should be assessed when specific energy use is the main consideration.

Although the RO process is relatively mature, opportunities exist to further reduce the energy use by small but economically significant amounts. These opportunities include the following:

- **Reduced fouling through pretreatment,**
- **Development of fouling-resistant membranes,**
- **Development of high-flux (i.e., high-permeability) membranes for operation at low pressure;**
- **Development of oxidant-resistant membranes; and**
- **Improving mechanical configuration of membrane modules and membrane system design.**

Operating the RO process at lower hydraulic pressure while maintaining high throughput is the key to reducing the specific energy for membrane-based desalination. To fully utilize the capacity of the high-permeability RO membranes and to accommodate even more permeable RO membranes in the future, it is imperative to reduce fouling and concentration polarization effects and to develop new module configurations and system designs to avoid or overcome thermodynamic restriction. Fouling can be reduced by a more robust pretreatment and by the development of fouling-resistant membranes.

Pretreatment for RO desalination can be improved by replacing conventional physicochemical processes with membrane-based (UF

or MF) pretreatment. Conventional pretreatment technologies based on coagulation and sand filtration cannot always achieve sufficient removal of foulants. Membrane-based pretreatment, particularly UF, can produce water of superior quality with very low fouling potential. Such effective pretreatment is essential for efficient utilization of future high-flux membranes.

Seawater desalination using thermal processes can be cost-effective when waste or low-grade heat is utilized effectively. Location of low-grade or waste heat resources near large water consumers may reduce the cost of heat energy and offset the higher specific energy requirements of thermal desalination when compared to RO. Hybrid membrane-thermal desalination approaches offer additional operational flexibility and opportunities for water production cost savings for facilities co-located with power plants. Thermal desalination technologies are themselves relatively mature; however, additional cost savings could be realized by improvements in materials, process configuration, and optimization of low-grade and waste heat resources.

Few, if any, cost-effective environmentally sustainable concentrate management technologies have been developed for inland desalination facilities. Several methods are currently available for concentrate management (e.g., surface water discharge, sewer discharge, deep-well injection, evaporation ponds, land application, and high-recovery/thermal evaporation systems to minimize the volume of waste produced), and each method has its own set of site-specific costs, benefits, regulatory requirements, environmental impacts, and limitations. Low- to moderate-cost inland disposal options can be limited by the salinity of the concentrate and by location and climate factors. Only evaporation ponds and high-recovery/thermal evaporation systems are ZLD solutions, but their costs are high for municipal application.

5

Environmental Issues

Desalination has been used around the world on a municipal scale for many decades, yet it is still considered by many to be a "new" option for addressing water supply needs. Part of the hesitancy to accept this technology comes from concerns over potential environmental impacts of desalination, which have not been fully quantified. The environmental issues surrounding desalination fall into four general categories, which are reviewed in this chapter: (1) impacts from the acquisition of source water, (2) impacts from the management of waste products and concentrate from the desalination process, (3) issues with desalinated product waters, and (4) the impacts of greenhouse gas emissions from these energy-intensive processes. Technologies and approaches to mitigate these impacts are also discussed. Environmental impact assessments for any project also include concerns that are not addressed in this chapter, such as environmental effects of plant construction, material use, potential releases to the air, disposal of used membranes, and socioeconomic considerations. These issues are discussed in a recent World Health Organization report "Desalination for Safe Water Supply" (WHO, 2007). Human use of any water supply will have some environmental impacts; ultimately, consideration of the potential impacts of desalination will need to be weighed against the impacts from other water supply alternatives.

SOURCE WATER ACQUISITION

Desalination technologies can provide high-quality water tailored to the user's needs, and many otherwise unusable sources of water (e.g., oceans, estuaries, brackish aquifers, wastewater) can be treated by desalination technologies. For each type of source water, there are distinct environmental considerations when that water is withdrawn. In coastal surface waters, issues of impingement and entrainment of marine organisms are paramount. For inland aquifer systems, the renewability of the resource and land subsidence over time are significant issues.

Marine Water Intake Issues: Impingement and Entrainment

Pumps bringing large volumes of ocean or estuary water into desalination plants can cause impingement and entrainment. Impingement, defined as the pinning and trapping of fish or other larger organisms against the screens of the intake structures, can cause severe injury and death to organisms. Entrainment occurs when intake pipes take in small aquatic organisms, including plankton, fish eggs, and larvae, with the intake water. Organisms that are pulled into the system will die if they are subjected to high temperatures or are crushed by high-pressure membranes. Intakes for desalination plants co-located with power plants are regulated under Section 316B of the Clean Water Act, although states may choose to apply these regulations to stand-alone plants as well (see Box 5-1).

Power plants have been well studied with regard to impingement and entrainment of organisms. Most desalination plants will take in far less water, roughly an order of magnitude lower than medium-sized power plants. However, very large stand-alone desalination plants might require comparable quantities of intake water if substantial volumes of water are needed for concentrate dilution. Intakes from a single large power plant are estimated to kill billions of juvenile-stage fishes each year and may affect recruitment of juvenile fish and invertebrates into the adult populations (Brining et al., 1981). It has been estimated that the magnitude of loss from one large power plant is equivalent to the loss of biological productivity of thousands of acres of habitat (York and Foster, 2005). The decomposition of the dead organisms can reduce the oxygen in the water, causing an additional stress in the area. However, the population-level impacts of mortality due to entrainment of marine organisms may or may not be substantial because the normal mortality of larval organisms in the marine environment is generally very high. The impacts and the acceptability of this loss will likely vary from place to place.

There are technologies and practices that can be applied to reduce the amount of impingement and entrainment associated with coastal desalination. To reduce the amount of entrainment, it is possible to reduce intake during the times when eggs and larvae are abundant in the water, and windows of operation can be set to minimize the entrainment of eggs and larvae of the species of concern. If intake pipes are located in deeper parts of a bay, there will also be fewer organisms that could be impinged or entrained (San Francisco Bay Conservation and Development Commission, 2005). Entrainment can also be reduced substantially by

BOX 5-1
Environmental Regulatory Framework for Desalination

Several national regulations serve as the legal framework to minimize environmental impacts from desalination processes. The most pertinent regulations are associated with the Clean Water Act, although the Safe Drinking Water Act and its Underground Injection Control program are also described here. Additional environmental regulations that need to be considered in the permit process are described in Chapter 7.

Clean Water Act

Under Section 316(b) of the Clean Water Act (CWA; P.L. 92-500), the Environmental Protection Agency (EPA) has developed regulations that require that the location, design, construction, and capacity of cooling water intake structures reflect the best technology available for minimizing adverse environmental impacts. Phase 1, promulgated in 2001, addresses the intake structures of new power plants. Phase 2 addresses the intake structures of large existing electric generating plants and requires these plants to meet impingement and entrainment standards for reducing the number of organisms affected. As of July 2007, the Phase 2 regulations were suspended while the EPA addresses several issues remanded by the 2nd Circuit Court of Appeals.[1]

In the United States, effluent discharges are federally regulated by the CWA. The regulatory program includes the National Pollutant Discharge Elimination System (NPDES), established by Section 402 of the CWA, which sets limits on the amount of various pollutants that a point source (i.e., the desalination plant) can discharge into a surface water body in a specific time period. Effluent limits can be technology based or water quality based, but they are all performance standards—that is, the permittee is free to use any combination of process modification, end-of-pipe treatment, or other strategies to meet them. Water quality standards can also vary depending on the specified use of the particular water body into which the concentrate is disposed. NPDES permits typically quantify areas—termed mixing zones—where surface waters may exceed water quality standards due to point source discharges.

If state regulatory programs meet the EPA requirements, the programs can be delegated to be administered by the states; therefore, regulations may vary somewhat from state to state. Effluent limits for desalination plants may specify pH, metaphosphates, chlorides, dissolved oxygen, conductivity, copper, iron, radium, total dissolved solids (TDS), total nitrogen, sulfide, ammonia, turbidity, radionuclides, selenium, and others. EPA (or the delegated state regulatory program) specifies the monitoring requirements, frequency of testing, and reporting, and the monitored species can be regionally variable. Some states require whole effluent tests for desalination concentrate in addition to chemical-specific numerical limits (Mickley et al., 1993). Whole effluent toxicity testing may include acute tests of 96 hours' duration using larval or juvenile fish and invertebrates, with survival as the end point, and chronic tests of 7 days in duration using early life stages of a fish and an invertebrate, considering metrics

continued

---

[1] For more information, see http://www.epa.gov/waterscience/316b/index.html.

such as growth. Local species may also be used instead of "standard" bioassay organisms if a bioassay has been developed for them and is approved by EPA.

Safe Drinking Water Act

The Safe Drinking Water Act (SDWA) regulates the levels of contaminants permitted in drinking water supplies and applies to every public water system in the United States. Based on data describing how often a particular contaminant occurs in the environment, how humans are exposed to it, and potential health effects of exposure, EPA sets a maximum contaminant level goal (MCLG), the level of a contaminant in drinking water below which there is no known or expected health risk, including a margin of safety. These goals are not enforceable because they do not take available technology into consideration and are sometimes set at levels that public water systems cannot meet. EPA proposes an enforceable standard in the form of a maximum contaminant level (MCL), which is the maximum amount of a contaminant allowed in water delivered to a user of a public water system. Every 5 years, EPA establishes a list of contaminants that are known or anticipated to occur in public water systems and may require future regulation under SDWA.[2]

EPA oversees deep-well injection of desalination concentrate through its underground injection control program within the SDWA. EPA has developed the following classification for injection wells (EPA, 2007b):

- Class I: wells that inject hazardous waste;
- Class II: wells associated with the oil and gas industry;
- Class III: wells that inject fluids for the extraction of minerals;
- Class IV: wells that inject hazardous or radioactive waste into a formation within one-quarter mile of a drinking water source; and
- Class V: all other injection wells not covered by Classes I-IV.

These classifications each have associated standards and associated regulations. States, rather than EPA, generally enforce the program and issue permits. Subsurface injection of desalination concentrates are covered by the states under regulations for either Class I or V injection wells.

reducing water intake volumes. Reducing the size of mesh of screens in intakes can reduce entrainment but will increase impingement. However, rotating screens and other types of technologies can minimize the intake of aquatic organisms. If intakes are placed below the surface through the use of beach wells or other subsurface intakes (see Chapter 4), the problems of entrainment of marine organisms are largely eliminated. Technologies for reducing impingement and entrainment are discussed in detail in Chapter 4.

Co-location of a desalination plant with an existing power plant takes advantage of existing intake structures (see also co-location in Chapter 7). Typically, a co-located desalination plant takes its source water from

[2] http://www.epa.gov/safewater/sdwa/30th/factsheets/standard.html#4.

the power plant discharge water; thus, as long as the power plant is operating, the desalination facility does not increase the impacts from impingement and entrainment. However, should the power plant discontinue operating on an interim or permanent basis or if once-through cooling practices are phased out, water withdrawals would have to continue to provide source water to the desalination plant. It is worth noting that many of the nation's power plants were sited decades ago, before the adverse environmental impacts of their intake structures were understood and before many of the current federal environmental legislation and regulations were in place, and some of the existing power plant intakes are located in areas where they create considerable environmental damage. Thus, the potential source water impacts of co-located desalination facilities still need to be considered.

Brackish Groundwater Source Issues

Some inland and coastal communities utilize brackish groundwater as a source for desalination, and withdrawal of brackish groundwater creates a quite different set of environmental concerns, including the physical sustainability of the aquifer and the potential for subsidence. Following a brief overview of brackish water resources in the United States, the potential environmental impacts from brackish groundwater withdrawal are discussed in more detail.

A brackish aquifer is a geologic deposit of water-bearing permeable rock or unconsolidated materials from which brackish groundwater can be usefully extracted using a well. The processes that generate brackish groundwater depend on the site-specific hydrogeology and geochemistry. In some cases, high levels of dissolved solids are derived from the presence of connate water (i.e., seawater trapped at the time of original deposition), but in most inland brackish water systems these original solutes have long since been flushed away. In arid and semi-arid areas typical of the western United States, the major sources of salinity in groundwater are the evaporative concentration of solutes from precipitation and dissolution of minerals in the subsurface. In the humid east and other areas with higher groundwater recharge rates, major solutes in brackish waters originate from dissolution reactions of the water with minerals (e.g., halite [NaCl], gypsum [$CaSO_4$], anhydrite [$CaSO_4 \cdot 2H_2O$], calcite [$CaCO_3$], dolomite [$CaMg(CO_3)_2$]) present in the aquifer framework.

Coastal aquifers form another class of natural brackish water created from mixing of groundwater that is discharging to the ocean. Under natural conditions most groundwater in coastal areas discharges directly to the ocean (Figure 5-1). The processes of molecular diffusion and hydrodynamic dispersion (mixing by movement of fluids through a porous media) create a brackish zone of dispersion or a mixing zone. Coastal groundwater pumping can cause seawater intrusion that increases the thickness of the brackish water zone of dispersion.

Brackish water from irrigation return flows can also be utilized as desalination source water, although the quantity and quality typically vary by season and region. In Colorado, some desalination plants use alluvial groundwater with elevated salinity as a result of agricultural land use in the drainage basin (e.g., Platte River). Development of this water source for desalination is site specific as to both quantity and quality.

Brackish groundwater exists at less than 305 m (1,000 feet) across much of the United States (Feth, 1965) (see Figure 1-1). This groundwater consists of highly variable concentrations of dissolved solids and ranges from slightly brackish to brines with salt concentrations many times the concentration of seawater. The distribution, volume, and water quality of brackish water aquifers in the United States are largely unknown. Some states, such as New Mexico (Huff, 2004a, 2004b; New Mexico Office of the State Engineer, 2004) and Texas (Brackish Groundwater Manual, 2003), have made brackish water inventories based on existing data. Huff (2004a) estimated that huge quantities of brackish water (16 trillion m^3 or 13 billion acre-feet) exist in New Mexico relatively close to the surface, some fraction of which could be desalinated for human use. However, there have been no national assessments, and the current regional assessments exhibit inadequate detail

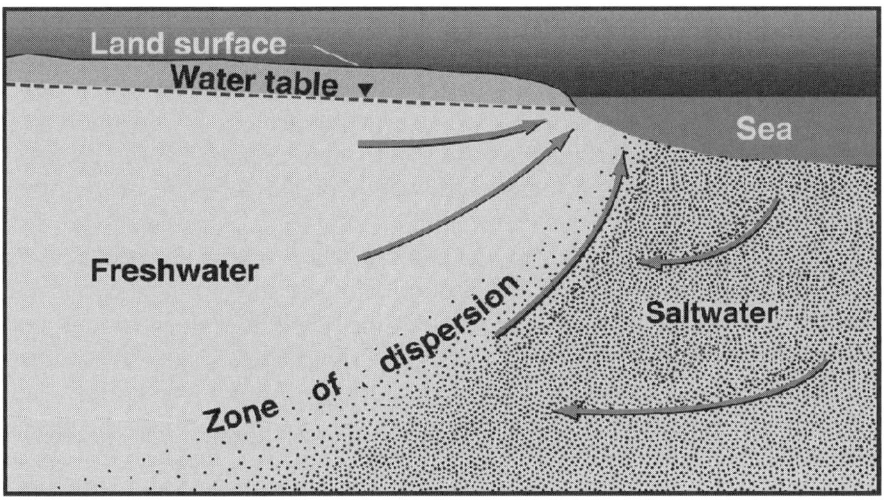

FIGURE 5-1. Diagram showing typical discharge of potable groundwater to the ocean and zone of dispersion (mixing zone).
SOURCE: USGS (2007a).

necessary for water resource management. Detailed site-specific evaluations, such as those conducted by the El Paso and Fort Bliss desalination facilities (see Box 5-2), will be necessary to assess the quantity and quality of water available for a given desalination facility. Nevertheless, a national compilation of existing data and regional evaluations of flow and solute boundary conditions, thickness, extent, and hydraulic conductivity of major brackish aquifer systems in the United States could provide the framework for a potentially greater utilization of brackish groundwater resources.

Physical Sustainability

Development of a brackish aquifer system for water supply demands an understanding the sustainability and renewability of the aquifer, in terms of both water quality and water quantity. The concept of physical sustainability of a natural resource has been defined in many ways. The United Nations Brundtland Commission (1987) popularized the term in the environmental sense when it defined sustainability as "the ability to meet the needs of the present generation without compromising the ability of future generations to meet their needs." For the present report, a more conservative viewpoint is taken; a physically sustainable aquifer system is considered to be one in which recharge over human time frames approximately equals withdrawals and discharges from both anthropogenic and natural processes (i.e., renewability). Groundwater withdrawals that exceed the recharge capacity of the aquifer are sometimes referred to as groundwater mining. Under these circumstances, continued withdrawals may deplete the groundwater resource, create subsidence (discussed below), or affect the quality and quantity of adjacent water bodies or aquifers. Because the hydrology of groundwater, lakes, streams, and wetlands are frequently interconnected, the removal of water from one source means less water for one or more of the other sources.

In terms of water quality, sustainable aquifers are defined here as those having concentrations that do not change significantly beyond the natural variability over human time frames. Solute concentrations and their ionic ratios are naturally variable throughout all aquifer systems (Hem, 1986), and increased pumping can induce groundwater flow and, thus, solutes from adjacent, underlying, and overlying aquifers or surface waters. Induced groundwater flow will, in most cases, lead to changes in water quality due to chemical reactions and transformations within the aquifer matrix (e.g., ion exchange, dissolution, precipitation). Mineral precipitation and dissolution in the aquifer matrix can potentially alter the hydraulic conductivity of the aquifer over time (Johnson et al., 2005).

Box 5-2
Kay Bailey Hutchison Desalination Plant

A desalination project was proposed for the El Paso, Texas, area after the 50-year Texas State-Wide Water Resource Management Plan for Far West Texas Region indicated that the projected future population growth of the El Paso area would experience water demands in excess of supplies. This proposed shortfall would occur in spite of the already large consumption decline associated with conservation and rate-structure change practices that resulted in a per capita decline from 0.870 m^3/day (230 gallons per day) in 1977 to 0.518 m^3/day (137 gallons per day) in 2006. At the same time the nearby U.S. Army facility at Fort Bliss was expanding its mission and sought to increase its water supply. Looking at all possible water sources, El Paso Water Utilities (EPWU) studies indicated that desalinated water would be less costly than using reclaimed sewage or importing additional water but approximately twice as expensive as current groundwater and surface water supplies. A proposal was made for a cooperative effort between the city of El Paso and the U.S. Army to develop a 10,400 m^3/day (27.5 million-gallon-per-day) desalination facility using brackish groundwater from the Hueco Bolson. The capital cost of the source water, desalination facility, and management of by-product concentrate was to be $87 million (2005 U.S. dollars).

Previous pilot plant operation by EPWU in 1993-1994 suggested the likely success of using reverse osmosis technology with Hueco Bolson water. The source water would have TDS ranging between 1,200 and 1,500 mg/L TDS and produce a final blended water of 700 to 800 mg/L TDS, comparable to existing water quality. The desalination facility would require 17 wells for source water and 16 wells for blending. The addition of antiscalant and mineral acid to the source water is designed to inhibit mineral precipitation on the membranes, transmission pipes, and walls of the injection wells.

A MODFLOW numerical model was constructed with cooperation by the U.S. Geological Survey and Fort Bliss, the International Boundary and Water Commission, and the City of Juarez Mexico Water Utility to evaluate the influence of removing brackish water on the aquifer system. This peer-reviewed model found that the relatively low projected rate of withdrawal for the desalination and blending from this large aquifer system would be environmentally benign and was sustainable through the projected life of the facility. Disposal of by-product fluids by well injection was determined to be cost-effective compared to both passive and enhanced evaporation procedures. Geophysical investigations and test wells indicated the Fusselman dolomite, a relatively high-permeability formation located 29 miles northeast of the desalination facility, would be hydrologically suitable for disposal. A disposal permit was obtained from the State of Texas to inject desalination effluent into this aquifer. The injected water was lower in TDS than the native water in the Fusselman formation and only exceeded drinking water standards for arsenic and selenium.

SOURCE: Ed Archuleta, EPWU, personal communication, 2006.

Ion exchange in certain instances may also change the physical properties of the ion-exchanger clay mineral and result in loss of permeability of the aquifer (Civan, 2000).

These impacts to water quality and quantity can be anticipated through hydrogeologic and geochemical analyses and the use of model simulations (see Box 5-3). The challenge to hydrogeologists is to identify and quantify all hydrologic boundaries and transport conditions under different scenarios of water supply development as well as any chemical reactions that add, subtract, or change solutes or ratios under varying flow conditions. Each well field is unique with respect to these properties and requires site-specific information.

Subsidence

When groundwater withdrawals exceed recharge rates, the hydraulic head (related to fluid pressure) will gradually be reduced and may result in land subsidence in areas with unconsolidated sediments (Figure 5-2). Subsidence results from the compression of the skeletal framework caused by reduced hydraulic head and rearrangement of grains in the aquifer matrix. Under equilibrium conditions, total stress (a function of the mass of water and rock) acting downward on a plane is balanced by a combination of the geologic framework acting upward (the skeletal framework resisting compression, or effective stress) and fluid pressure (hydraulic head). The reduction in hydraulic head associated with withdrawal of fluids increases the effective stress on the geologic framework and causes compaction and reduction in elevation of the land surface (Tergazi and Peck, 1967). Excessive development of a brackish water resource, particularly one that possesses thick sections of at-risk lithologies (i.e., clay, silt, organic material), may create the potential for subsidence.

Land subsidence resulting from removal of groundwater has affected areas in 45 states (Figure 5-3) and ranges from regional lowering to ground failure and collapse (Galloway et al., 1999; NRC, 1991). Galloway et al. (1999) estimate that more than 80 percent of the identified subsidence in the United States has been caused by overexploitation of groundwater resources. Parts of Texas, California, and Nevada have experienced tens of meters of surface decline (Leake, 2007). Regional lowering will increase the probability of flooding in coastal areas, while local settling may damage engineered structures such as buildings, roads, and utilities. Subsidence also has the potential to trigger earthquakes and activate faults; for example, the Gulf Coast basin is a region where subsidence and fault activation are common around large, mature oil and gas fields (USGS, 2007b).

BOX 5-3
Tools for Addressing Brackish Groundwater Physical Renewability

Several modeling tools are available to assess the potential impacts of a proposed brackish groundwater desalination facility on the physical renewability of an aquifer. Once appropriate site-specific information is obtained, numerical groundwater models, such as variations of MODFLOW (Harbaugh, 2005), can be used to test various scenarios of development and effects on water quantity. If density contrasts between water bodies are significant, then codes such as SEAWAT-2000 (Langevin et al., 2003) and SUTRA (Voss and Provost, 2003) may be more appropriate codes to use. Other models, such as PHRQPITZ (Plummer et al., 1988) and PHREEQC (Parkerhurst, 1995), address water quality impacts.

MODFLOW-2005 simulates steady and nonsteady flow in an irregularly shaped flow system in which aquifer layers can be confined, unconfined, or a combination of confined and unconfined. Flow from external stresses, such as flow to wells, areal recharge, evapotranspiration, flow to drains, and flow through riverbeds, can be simulated. Specified head and specified flux boundaries can be simulated. In addition to simulating groundwater flow, MODFLOW-2005 incorporates related capabilities such as solute transport and groundwater management. SEAWAT-2000 is the latest release of the SEAWAT computer program for simulation of three-dimensional, variable-density, transient groundwater flow in porous media. SUTRA is a model for saturated or unsaturated, variable-density groundwater flow with solute or energy transport. SUTRA version 2D3D.1 includes both two- and three-dimensional simulation capability.

Water quality changes and aquifer transformation in brackish or brine conditions can be assessed using the model PHRQPITZ, which addresses chemical reactions (e.g., mineral precipitation, dissolution) within brackish aquifers. PHRQPITZ is a computer code that permits calculations of geochemical reactions in brines and other highly concentrated electrolyte solutions using the Pitzer virial-coefficient approach for activity-coefficient corrections. Reaction-modeling capabilities include calculation of (1) aqueous speciation and mineral-saturation index, (2) mineral solubility, (3) mixing of aqueous solutions, (4) irreversible reactions and mineral-water mass transfer, and (5) reaction path.

PHREEQC (version 2), which adds the Pitzer coefficients for dealing with brackish water, is a computer program that is designed to perform a wide variety of low-temperature aqueous geochemical calculations. PHREEQC has capabilities for (1) speciation and saturation-index calculations, (2) batch-reaction and one-dimensional (1D) transport calculations, and (3) inverse modeling. Transport capabilities involve reversible reactions (e.g., aqueous, mineral, gas, solid-solution, surface-complexation, and ion-exchange equilibria) and irreversible reactions (e.g., kinetically controlled reactions, mixing of solutions, temperature changes). PHREEQC version 2 includes new capabilities to simulate dispersion (or diffusion) and stagnant zones in 1D-transport calculations and to model kinetic reactions with user-defined rate expressions.[3]

[3] For more information on these models, see http://water.usgs.gov/software/lists/ground_water/.

FIGURE 5-2. Land subsidence from groundwater withdrawal.
SOURCE: Galloway et al. (1999).

FIGURE 5-3. Areas where subsidence has been attributed to groundwater withdrawal.
SOURCE: Galloway et al. (1999).

Potential of Brackish Groundwater to Meet Future Water Needs

The volume of inland brackish groundwater in the United States is unknown; based on the relative shallow depths of brackish groundwater reported by Feth (1965) (see Figure 1-1) and surveys of several states in the southwestern United States (Huff, 2004a, 2004b), large quantities of brackish water appear to be available for development. However, knowledge of the brackish aquifer systems in the United States is insufficient to assess whether they can be developed without long-term groundwater mining or water quality deterioration of the primary or adjacent aquifers. Additional detailed studies will be needed to assess the potential impacts of increased groundwater withdrawals at any specific site.

CONCENTRATE MANAGEMENT

Desalination processes create waste products, including salt concentrates, cleaning and conditioning reagents, and particulate matter, that must either be disposed of or reused (Malmrose et al., 2004). This section will focus on the waste products unique to seawater or brackish groundwater desalination and potential environmental impacts from common concentrate management practices. It is worth noting here that the limited research to date on the environmental effects of concentrate management practices has primarily been focused on seawater desalination plants, because globally, seawater plants tend to be the largest desalination facilities. Currently, the majority of desalination facilities in the United States, both in the number of plants and in total capacity, use brackish source water (see Chapter 2).

Chemical Constituents in Desalination Concentrate

The largest component of the waste stream from a desalination facility is concentrate that varies in salt concentration and in the ratios of particular species with the specific source water (see Table 2-1) and the desalination process used. The concentrate stream from a desalination process includes the constituents rejected from the feedwater stream, in a more concentrated form. Currently, depending on the source water quality and constituents, the recovery of brackish water membrane desalination is between 50 and 90 percent, implying that 10 to 50 percent of the feedwater is converted to concentrate (concentrating the solutes by a factor of 2 to 10). For seawater desalination, the corresponding recovery range is 35 to 60 percent, implying that 40 to 65 percent of the feedwater is converted to concentrate (concentrating the solutes by a factor of 1.5 to 2.5). In addition to salts, other naturally occurring elements from the source water are concentrated (e.g., selenium from some brackish groundwater sources), which, depending on the source water quality and discharge concentrations, could cause adverse impacts if discharged into sensitive environments. When desalination plants are co-located with power plants, desalination concentrate may also contain excess heat that may pose concerns for disposal into the environment. Desalination concentrate may also contain chemicals used in the desalination process (see Table 4-1). The nature and concentration of the chemicals in the concentrate vary and are site-specific, depending on which chemicals are used, the amounts used, how frequently they are used, and whether they are discharged in the concentrate or disposed of via sewage treatment plants or other disposal options. The nature and potential impacts of these chemical additives are discussed in detail below.

Biocides, such as sodium hypochlorite for chlorination, may be added to minimize fouling in surface water (seawater and surface brackish water) desalination (Einav et al., 2002). Chlorine is quite toxic to marine biota if discharged to the environment without neutralization. Because EPA criteria for short- and long-term exposure are 13 and 7.5 μg/L, respectively (EPA, 1985), and because there is potential for creating environmentally persistent, bioaccumulative, and toxic halogenated organic compounds after chlorine addition, chlorine is considered a serious concern (Lattemann and Höpner, 2003). Lattemann and Höpner (2003) report that chlorine has been found at concentrations up to 100 μg/L within 1 km of desalination plant outfalls in Kuwait, which uses primarily thermal desalination plants, usually co-located with power plants. However, the high chlorine levels at the Kuwait plants would likely continue even if the desalination plant were no longer operating because of the biocide requirements of the power plants. Chlorination is not a major issue for most reverse osmosis (RO) plants—the vast majority of the desalination facilities in the United States. Chlorination in the RO process is typically intermittent, and the added chlorine is neutralized because membranes are sensitive to oxidizing chemicals.

Coagulants, such as ferrous chloride and aluminum chloride, are often added in seawater and surface brackish water pretreatment processes to remove suspended matter. These chemicals, released at 1-30 ppm in the concentrate, are of very low toxicity but can form precipitates and may cause increases in turbidity and discoloration in the case of iron compounds (Lattemann and Höpner, 2007).

Filters in RO plants need to be backwashed every few days to clear the accumulation of solids. In the United States, this filter backwash is not permitted to be directly discharged to the environment, because it can cause both considerable discoloration in the water at the discharge site and beach contamination. However, the practice may occur in other locations.

Antiscaling substances, antifoaming additives, oxygen scavengers, and anticorrosion chemicals may also be present in the discharge of seawater concentrate (Höpner, 1999; Rachid and Abdelwahab, 2005). Antiscalants, which are added to inhibit the formation of scale precipitates and salt deposits, belong to different chemical groups. Commonly used antiscalants are polyphosphates, phosphonates, or carboxylic-rich polymers such as polyacrylic acid and polymaleic acid. Polyphosphates are increasingly being replaced by the other polymers that are more stable at high operating temperatures and more resistant to chemical and biological breakdown. All antiscalants are nontoxic at the levels used for desalination (1-2 ppm in the concentrate). However, issues that may be of concern are nutrient levels in the discharge, which may be increased by polyphosphate, and the relatively slow degradability of the other antiscalant species (Lattemann and Höpner, 2003, 2007). Little is known about environmental effects of the polyphosphate scale control additive sodium hexametaphosphate, which was one of the first antiscalants to be used, although it is known to act as a nutrient. Polymer antiscalants such as polyacrylamide, however, are of low toxicity and low bioaccumulation and might actually reduce copper toxicity by binding to it (Lattemann and Höpner, 2003).

Antifoaming additives include acylated polyglycols, fatty acids, and fatty acid esters, which are detergents (Höpner, 1999). Polyethylene glycol (PEG) is nontoxic and is used in low concentrations, but it is persistent in the environment. PEG is specifically used in the multistage flash (MSF; see thermal processes in Chapter 4) process and usually only temporarily when natural surfactants cause seasonal foaming. Some MSF units with aggressive process designs may be more sensitive to foaming than others that may never require antifoaming agents (Lattemann and Höpner, 2003). Benzotriazole derivatives may be used as corrosion inhibitors and may be released during the cleaning of MSF plants. Benzotriazole chemicals are persistent, but have low potential to bioaccumulate and are toxic only at very high concentrations.

Desalination concentrate from any source water may also contain metals from corrosion that may be toxic to organisms if they are discharged into the environment. Concentrations of metals in the desalination discharge will vary with the design and the specific materials used in the desalination plant. Chemicals of corrosion are unlikely to be a major concern for RO plants, although RO plants will discharge minor amounts of iron, chromium, nickel, and molybdenum in their concentrate from stainless steel (Lattemann and Höpner, 2007). Chromium in its hexavalent form is known to be quite toxic, but there has been very little study of metal toxicity from RO plants.

Corrosion-related chemicals are primarily an issue with MSF plants, because sections of the MSF process operate at higher, corrosive temperatures. Copper is estimated at 5-33 μg/L in the blended discharge of co-located MSF facilities, which is above the EPA criteria of 4.8 and 3.1 μg/L for short- and long-term exposure, respectively (EPA, 1985; Lattemann and Höpner, 2003). While dissolved copper can form nontoxic complexes and bind to sediments, chronic effects are possible. Copper is an essential metal, but some organisms, such as algae and mollusks, are extremely sensitive to elevated levels. The most toxic form of copper is the free ion (Cu^{2+}), which tends to be of low concentration in the water column because most of the copper binds to suspended material and accumulates in the sediments. However, sediment-bound metals can be remobilized by a sudden change in environmental conditions and serve as a source of metal contamination to benthic and other marine organisms. Keeping the concentrate pH higher and adding chelators can reduce the concentration of free copper and, thus, the toxicity of the concentrate (Rinne, 1971). Nickel is less toxic than copper (EPA criteria are 74 and 8.2 μg/L for short- and long-term exposure, respectively [EPA, 1986]), and although its concentration in the effluent may be similar to that of copper, it is not considered a serious risk. MSF plants built in the past 10 years have become much more corrosion-resistant, because titanium has increasingly been used for the tubes in the hotter sections of the plants. However, this trend may have halted or reversed due to the increased cost of titanium. Projects under construction now may once again rely primarily on copper alloy tubes (Lattemann and Höpner, 2003).

Chemicals such as detergents are also used periodically to clean desalination membranes to improve system performance, prevent membrane fouling, and extend membrane life (see Table 5-1). These may or may not be discharged with the concentrate. Acid is typically used to remove inorganic foulants (e.g., metal oxides or scale), and alkaline solutions and dispersants are used to remove biofilms, silt deposits, and organic foulants. Thus, membranes are cleaned with the use of low pH, high pH, oxidant, biocide (e.g., formaldehyde), and detergent solutions. If acidic detergents are used and not disposed separately, discharge of the concentrate into freshwater may have impacts on the pH of the receiving water, even if diluted. Seawater tends to be well buffered and is not likely to experience significant acidification. These chemicals can also be considerably toxic to freshwater and marine organisms. The nature and volume of cleaning solutions and waste depend on the system design, the size of the system, and the frequency of cleaning. Most plants clean their membranes every 3 months, or less frequently. Typically the volumes of cleaning solutions used (see Table 5-1) per cleaning are about 1.2 L/m^2 (3 gallons per 100 square feet) of membrane (AWWA Membrane Residuals Management Subcommittee, 2004).

TABLE 5-1 Typical nanofiltration and reverse osmosis cleaning formulations.

Foulant Type	**Cleaning Solutions**
Inorganic salts[a]	0.2% HCl
	0.5% H_3PO_4
	2% citric acid
Metal oxides	2% citric acid
	1% $Na_2S_2O_4$
Inorganic colloids (silt)	0.1% NaOH, 0.05% Na dodecyl benzene sulfonate, pH 12
Silica (and metal silicates)	Ammonium bifluoride
	0.1% NaOH, 0.05% Na dodecyl bezene sulfonate, pH 12
Biofilms and organics	Hypochlorite, hydrogen peroxide, 0.1% NaOH, 0.05% Na dodecyl bezene sulfonate, pH 12
	1% sodium tripolyphosphate, 1% trisodium phosphate, 1% sodium EDTA

[a]Barium sulfate, calcium carbonate, calcium sulfate.

SOURCE: Dow Chemical Co. (1994).

Lattemann and Höpner (2003) discussed the toxicity and potential risk of several common membrane cleaning chemicals, considering typical discharge concentrations and dilution, focusing on desalination plants in the Arabian Gulf. Sodium dodecyl benzene sulfonate (Na-DBS) was found to degrade relatively quickly (within a few weeks). There are no toxicity data for Na-DBS, but a similar detergent, sodium dodecylsulfate (Na-DSS), has moderate toxicity in fish, an invertebrate, and algae (LC_{50} or the concentration of salt lethal to 50 percent of the organisms of 1-10 mg/L). Na-DBS is designated as a hazardous substance under the Federal Water Pollution Control Act. In contrast, EDTA was found to degrade slowly (5 percent reduction in 3 weeks) but showed low bioaccumulation and low toxicity (acute effects at 10-100 mg/L). Sodium perborate, which was found to degrade at discharged concentrations, also has fairly low toxicity (LC_{50} of about 10-50 mg/L). Isothiazole at the levels used for membrane cleaning could be toxic in the immediate mixing area before dilution. Routine cleanings tend to be more benign and utilize organic acids, particularly citric acid, and facilities should perform neutralization prior to blending or metered discharges. Aggressive cleanings with EDTA or other chemicals are less common (Lattemann and Höpner, 2003). Lattemann and Hőpner (2003) conclude that membrane-cleaning chemicals can be quite hazardous, since they include disinfectants (e.g., formaldehyde, isothiozole), which interfere with cell membranes and are toxic at low concentrations.

These process chemicals and the suspended solids can, in most cases, be removed from the concentrate or neutralized rather than being discharged directly into the environment. In fact, most facilities in the United States discharge their cleaning solutions separately to sewers, but approximately 17 percent of RO facilities in the United States discharge their cleaning solutions with the concentrate (Kenna and Zander, 2000, 2001). The volume of chemical additives required may also be reduced by using various filtration methods in the pretreatment process (Dudek and Associates, 2005). Removing or neutralizing added chemicals before discharging the concentrate reduces the environmental impact.

In overall considerations of the contamination of the Arabian Gulf, where there are numerous MSF thermal and RO plants, Lattemann and Höpner (2003) considered copper and chlorine to be the most serious environmental threats from seawater desalination concentrate discharge.

Chlorine in regions of Kuwait Bay is present at levels close to toxic concentrations for some phytoplankton, invertebrates, and vertebrates. Höpner and Lattemann (2002) estimated that 21 thermal desalination plants released a total of 2,708 kg of chlorine, 36 kg of copper, and 9,478 kg of antiscalants daily directly into the Red Sea. (Comparable chemical discharge data for RO facilities are not available.) Copper is less likely to be a problem in the United States, because RO plants do not generally generate copper in significant amounts. Although dilution of the concentrate will reduce the environmental impacts of the higher salinity, it will not negate the effects of these other chemicals, because the overall loading—not the concentration—is a stress on the receiving environment (Höpner, 1999). Lattemann and Höpner (2003) also emphasize that nothing is known about the potential interactive effects of the various chemicals that may individually be present at nontoxic levels.

It is difficult to make broad assessments of the impacts of desalination process chemicals because the treatment approaches vary widely from plant to plant. In addition, some plants will discharge more chemicals in the concentrate whereas others will send the chemicals to sewage treatment plants or to other disposal options. The chemicals in RO concentrate tend to be less hazardous that those in thermal desalination concentrate, which can contain high levels of metals and chlorine. Nevertheless, membrane cleaning chemicals could have environmental impacts if they are discharged into the environment with the concentrate rather than being disposed of separately. Many of the pretreatment chemicals (e.g., antiscalants, coagulants) used in RO desalination are of relatively low toxicity, but there remains a need for more information on the environmental effects of the various chemicals that are used in the desalination process, both individually and especially in combination. Research should examine ways to reduce additives in concentrate discharge that can be harmful to the environment.

Environmental Impacts from Concentrate Management Approaches

As discussed in Chapter 4, concentrate management strategies may take place in a variety of ways, including surface water discharge (e.g., into oceans, seas, estuaries, lakes, rivers), wastewater discharge, injection into underlying aquifers, land application, evaporation ponds, and disposing of the salts in landfills after thermal evaporation. Various site-specific factors may limit the range of concentrate management options at a given site, such as state and local permitting, hydrological conditions, low season flows, capacity of local sewers, level of dissolved solids and toxic materials in the concentrate, availability of dilution water for land application, climate suitability, and costs of the different options (AWWA Subcommittee, 2004). These concentrate management approaches are implemented within the context of state and federal environmental regulations, as described in Box 5-1. These various concentrate management technologies and approaches and their relative use in U.S. desalination plants are described in Chapter 4 (see Figure 4-15). Potential environmental impacts from coastal and inland concentrate management methods, and approaches to mitigate those effects, are discussed in the following sections.

Coastal Concentrate Discharge into Oceans, Seas, and Estuaries

The salinity of seawater desalination concentrate can approach 2.5 times the salinity of seawater, and the impacts of this discharge on the marine environment will vary with discharge method, source water salinity and quality, site conditions, and ecosystem type. Without proper dilution, a plume of elevated salinity discharge may extend for a considerable distance beyond the mixing zone and can harm the ecosystem (Younos, 2005). If a desalination plant is co-located with a power plant or a wastewater treatment facility, the concentrate can be blended and diluted with the power plant discharge or treated wastewater effluent before being returned back to the environment, reducing the potential for salinity stress in organisms in the receiving water. In the case of co-location with a power plant, however, the blended effluent will be of higher temperature and can have temperature stress effects on local organisms. If concentrate is released in such a way as to maximize dispersion of the effluent, with diffusers and dense jets and nozzles arranged in particular configurations, these effects could be reduced. In addition to the configuration of the outlet, the extent of mixing will also depend on local site conditions, including the bathymetry, waves, currents, and depth of the water (Einav et al., 2002).

Concentrate from brackish water desalination plants may have salinities that are similar to or lower than seawater, depending on the source water salinity and the recovery efficiency. However, depending on the source water composition, concentrate from brackish groundwater desalination may also contain trace elements, such as arsenic or selenium, in elevated concentrations compared to seawater. Additionally, when concentrates originating from groundwater are discharged to seawater or brackish surface waters, major ion toxicity can result (Mickley et al., 2001). This toxicity occurs when certain ions are present in very different concentrations (higher or lower) relative to seawater diluted to the same salinity. Toxicity due to an "imbalance" of ions relative to seawater has been seen in mysid shrimp with respect to high calcium or fluoride or low potassium. This toxicity has sharp thresholds, but appropriate concentrate dilution should eliminate it. The toxicity in brackish groundwater membrane concentrate from high levels of hydrogen sulfide or low levels of dissolved oxygen can be remedied by aeration.

Different coastal and marine ecosystems are likely to vary in their sensitivities to concentrate discharge. Höpner and Windelberg (1996) ranked various marine ecosystems in terms of their perceived sensitivity to a seawater desalination plant's discharge. This ranking ranged from the least sensitive—high-energy oceanic coast and exposed rocky coasts—to the most sensitive—coral reef, salt marsh, and mangrove. However, the sensitivity scale they used was based on the sensitivity of different environments to oil spills. Thus, more study will be needed to determine the relative sensitivity of different types of estuarine and coastal environments to concentrate discharges. For example, although salt marshes and mangroves are very sensitive to the effects of oil spills because oil will persist in the sediments for a very long time, these estuarine sites will probably have a higher tolerance to increased salinity, because estuaries normally experience fluctuations in salinity. While Höpner and Windelberg (1996) considered high-energy rocky coasts to be relatively insensitive (as they are to oil spills), rocky habitats with kelp beds along the California coast are considered critical sensitive ecosystems (Cooley et al., 2006).

Studies of Effects of Concentrate on Specific Marine Organisms. There have been numerous papers discussing the potential for negative environmental impacts of effluents from desalination facilities, but there is a surprising paucity of useful experimental data, either from laboratory tests or from field monitoring, to assess these impacts. In sensitive organisms, increased salinity can cause osmotic pressure changes in cells and may affect respiration and photosynthesis and reduce growth rates (Rinne, 1971). However, different species have different levels of tolerance to increased salinity. Some organisms have very high salinity tolerances and are referred to as euryhaline, but others have only narrow tolerances and are called stenohaline. More euryhaline species are likely to be found in estuaries, where there is a natural variation in salinity every day with the tidal cycle, than in the ocean, where the salinity is more constant. In general, larvae and young individuals are more sensitive to environmental stresses than adults of the same species. The following section describes what is known from research about the potential impacts of seawater desalination concentrate on various marine biota.

The effects of concentrated salt on marine organisms have been tested in some laboratory bioassays. Pillard et al. (1999) exposed adults of three estuarine species (sheepshead minnow, *Cyprinodon variegatus*; mysid shrimp, *Mysidopsis bahia*; and silversides, *Menidia menidia*) to water of different salinities and monitored their survival over a 48-hour period (Figure 5-4). Whereas the very tolerant sheepshead minnow tolerated seawater of over 60 ppt, the mysids and silversides began to exhibit mortality at salinities of around 38-40 ppt. Based on this information, there have been conclusions that 38-40 ppt salinity represents a salinity tolerance threshold for marine organisms (Jenkins and Graham, 2006). While this estimate of salinity tolerance may indeed be correct, the data are not adequate to support it. These data reflect exposure over a very short time period and are focused only on mortality; thus, the data are far from adequate to establish what level of salinity is safe in the long term for marine organisms. Also, all three of these species are estuarine and very tolerant of changes in salinity.

Latorre (2005) reported on effects of concentrate on the seagrass *Posidonia oceanica*, which is the most widespread species of seagrass in the Mediterranean and is subject to discharges from an RO seawater desalination plant in Mazarron Bay, Spain. Mesocosm studies in the laboratory revealed that brief (15-day) exposures to 40 ppt salinity water caused 27 percent mortality and reduced the growth of surviving plants. At field sites near the RO plant, seagrass beds were degraded in areas where the salinity had increased to 39 ppt.

Early life stages tend to be more sensitive than adults. For the California grunion (*Leuresthes tenuis*), Reynolds et al. (1976) determined that prolarvae (i.e., larvae with a yolk sac, up to about 4 days old) have an upper salinity tolerance (LC_{50}) of 41 ppt after 24 hours of exposure and that 20- to 30-day-old larvae tolerate a maximum of 40 ppt for about

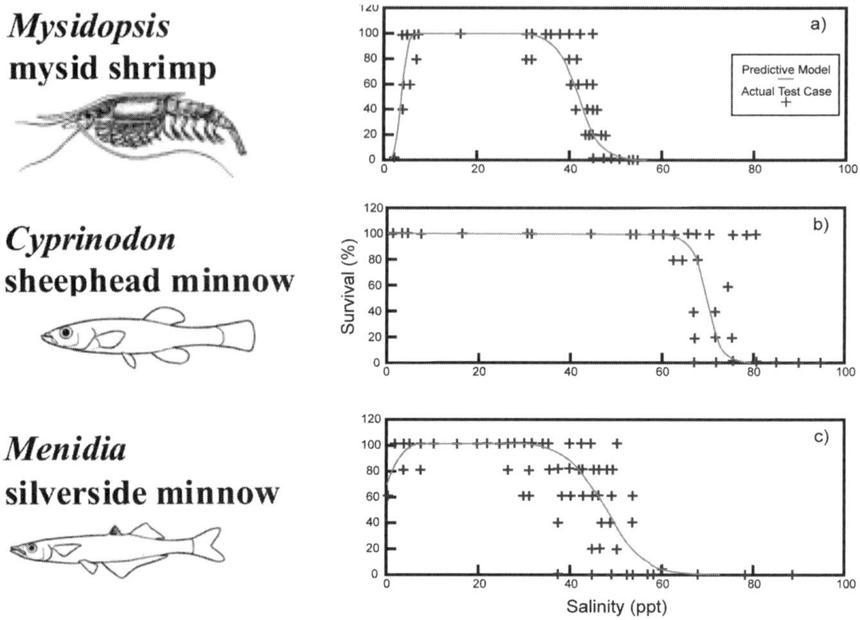

FIGURE 5-4. Whole effluent toxicity (WET) salinity tests on marine animals. Data reflect survival after 48 hours of continuous exposure to various salinity levels. SOURCES: Pillard et al. (1999), Jenkins and Graham (2006). Reprinted with permission from Environmental Toxicology & Chemistry, Allen Press Publishing Services.

18 hours. Lasker et al. (1972) found that salinities greater than 40 ppt adversely affected eggs and larvae of the California Sargo (*Anisotremus davidsoni*). Testing larvae is a good approach, but to determine adequate safety it is necessary to examine long-term sublethal effects and whole life cycles of various local species to see to what degree their physiology, behavior, growth, reproduction, and other parameters would be affected by different salinities. Studies on invertebrates have also mostly focused on short-term survival at increasing salinities and did not examine long-term or life-cycle effects (Bay and Greenstein, 1993; Blazkowski and Moreira, 1986; Charmentier and Charmentier-Daures, 1994; Ferraris et al., 1994; Forster, 1998).

A few studies have investigated sublethal effects of RO discharge water, but the effects and sensitivities differed from species to species. For example, Bay and Greenstein (1993) found that exposure of giant kelp (*Macrocystis*) blades to elevated salinities of 38.5 and 43 ppt did not affect either the spore generation rate or the length of the germination tube. A 14-day study of the California Sargo (*Anisotremus davidsoni*) found that the optimum salinity for juvenile feeding and growth was 33-

37 ppt, and adverse effects, such as reduced growth, were seen at 45 ppt (Brocksen and Cole, 1972). Iso et al. (1994) studied the effects of concentrate on eggs and larvae of fish species and found that salinities below the incipient lethal level could produce sublethal effects, such as retarded development and growth.

Field Studies of the Effects of Concentrate Discharge. Even after dilution, concentrate discharge will be of higher salinity than the receiving water and will tend to sink and settle along the bottom of the marine environment. Thus, the benthic community will be particularly impacted, although the specific effects will be dependent on the local ecosystem, the composition of the discharge, and the degree of dilution at the point of contact (Mickley et al., 2001). Perez Talavera and Quesada Ruiz (2001) studied impacts of the discharges in the Canary Islands, specifically examining effects on the seagrass *Cymodocea nodosa* and the alga *Caulerpa prolifera*. They found that although the initial dilution was high, salinities on the bottom remained elevated over a large area. The concentrate also elevated the turbidity of the water. Seagrasses were absent in areas near the outfall; farther away they were present but in poor condition and covered with slime. At distances even farther from the discharges, the seagrasses were in good condition.

A comprehensive study of the ecological impacts of concentrate discharge from a thermal desalination plant was performed by Chesher (1975) in Key West, Florida. This study examined chemical and physical properties of the discharge, historical analysis of sediments, and the abundance of benthic fauna over time. Laboratory bioassays were also performed on a number of species. Results indicated that the heated effluent, which was highly contaminated with dissolved copper, had a significant impact on the biota near the discharge. Temperature and salinity of the effluent were such that the effluent settled at the bottom of the receiving basin, which reduced water circulation. An 18-month biological study showed a marked reduction in biotic diversity.

Gacia et al. (2007) studied the effect of concentrate on a shallow seagrass meadow (*Posidonia oceanica*) exposed to RO concentrate discharge for over 6 years and compared the results with two undisturbed reference zones. The concentrate contained both elevated salinity and nitrogen because of high nutrient levels in the source groundwater—not a typical situation for desalination plants. Sea urchins and sea cucumbers—species very sensitive to salinity—were present in the reference transects but absent from the impacted transect. The seagrass proved to be sensitive to both elevated nutrients and salinity in the discharge; however, the observed necrosis and reduced growth in the seagrass and the absence of sea urchins were attributed to the high salinity. There was no indication of extensive decline of the affected meadow, probably due to its very shallow condition, which results in fast dilution and dispersion of the concentrate plume.

An intensive field study was performed to estimate the effects of concentrate discharge in Tampa Bay, Florida, using a study site on the island of Antigua under the assumption that it was similar to Tampa Bay (Blake et al., 1996; Hammond et al., 1998). The researchers redirected the Antigua seawater RO desalination plant's discharge to a new site, which they studied before and during exposure to the concentrate. This study ultimately found no effect of salinity elevation on the tropical reef ecosystem (see Box 5-4 for detailed findings), and it is the most comprehensive and thorough study that has been done to date, although it has not been published in the peer-reviewed literature. Despite the drawbacks of the study as described in Box 5-4, this extensive study should be a model for future studies in other locations.

In contrast to the minor changes seen in the previous study, Pilar Ruso et al. (2007) found major impacts on benthic communities following the 2003 opening of a seawater RO plant in Alicante, Spain, which discharges 65,000 m^3/day of 68 ppt TDS concentrate. The salinity at stations farther from the discharge was 37.9 ppt, while near the discharge it was elevated to over 39 ppt. Investigators studied benthic infaunal communities along three transects perpendicular to the coast. At the beginning of the study, the most abundant groups of organisms at all stations were polychaetes, nematodes, and bivalves. However, at 9 months from the beginning of the discharge, in the station closest to the discharge, the community became dominated by nematodes, and this dominance increased over time from 68 to 83 to 96 percent in subsequent samplings. During the 2 years after commencement of the discharge, other stations in the transect receiving the discharge showed a similar change in the structure of the community to becoming dominated by nematodes, reflecting an extension of the area of influence of the discharge. The authors recommend increasing the mixing of the concentrate, for example by diffusers, to reduce the area affected by high salinity.

In contrast, Raventos et al. (2006) found no significant impacts attributable to concentrate discharges from a small seawater desalination plant in the northwest Mediterranean on organisms present on the surface. Visual censuses were carried out 12 times before and 12 times after the plant began operating. No significant variations attributable to the discharges from the plant were found. The failure to observe impacts may be explained by the high natural variability and by the rapid dilution undergone by the concentrate upon leaving the discharge pipe, which had a diffuser with 43 perforations to facilitate rapid dilution of the discharge (salinity was back to normal 10 m away from the pipe).

BOX 5-4
Antigua Study of the Impacts of Concentrate Discharge

The study at Antigua was conducted to investigate the potential impacts of seawater desalination concentrate discharge in Tampa Bay. Prior to the concentrate discharge diversion to the study site, the researchers conducted baseline studies of the habitat. Six transects, extending out radially 10 m from the discharge site, and at 60-degree angles from one another, were established (see Figure 5-5). Transects extended both on- and offshore from the discharge site. The study area contained a diverse assemblage of organisms, including seagrass (*Thalassia*), algae (*Dictyota*), hard corals (*Porites*) and soft corals (*Pseudoterogorgia*), as well as other invertebrates and fishes. Results indicated that the discharge water had elevated temperature (2-3°C warmer), elevated salinity (see below), and a reduced pH (0.2-0.3 units lower) compared to the ambient water. Differences between the discharge and the ambient water were detectable within the study area but were rapidly dissipated by mixing. Dye injection demonstrated that the plume rapidly dissipated and moved toward deeper water. The elevated salinity "signal" was detectable beyond the 10 m study area and distributed mainly down slope. Maximum bottom salinities, recorded in the immediate vicinity of the discharge opening, were 35-40 ppt in June and 34-38 ppt in October; but because the pipe discharge flowed upward and contacted the surface, surface salinities were higher (35-44 ppt in June, 34-43 ppt in October). Because of strong mixing, salinities at the 8-10 m transect positions averaged only 0.2 ppt above ambient, with salinity increases extending farther downslope than upslope.

No significant changes were noted in biological communities along the plume. Studies of the seagrass beds indicated no changes in the number of new shoots, biomass, or growth rate over the survey period. All plants showed numerous parrotfish bites, which indicated that this fish frequented the study area in spite of the elevated salinity. A brown alga (*Dictyota*) showed variations in growth rate and a weak correlation with salinity. Tissues from plants living within the study area also showed a higher concentration of nitrogen than did plants sampled from outside the study area. Diatom numbers and types did not change from prediversion conditions. Benthic foraminifera distribution and abundance varied considerably and reflected a "stressed habitat," but there were no differences that related to the presence of the concentrate discharge. Dominant benthic infauna were annelids and one snail species. There were significant differences in the infaunal assemblage at different times (March and October had more animals than June), but abundance and diversity did not appear to be affected by elevated salinity. Epifauna were collected on settling plates. Bryozoans and polychaete worms were the dominant forms, with hydroids, snails, clams, and sea urchins also settling. Variation in the groups that settled on the plates at the different sampling periods was attributed to biological factors (reproductive season) rather than an elevated salinity. However, because there were no settling plate data prior to diversion, it is unknown whether or not increased salinity excluded any species from settling. Coral heads in the transect area exposed to an average salinity elevation of 4.5 ppt showed no ill effects over the 6-month observation period. There were no obvious effects of the discharge on either the macroinvertebrates or fishes in the different observation periods. However, no survey of fishes was done before the diverting the discharge.

continued

There are five notable limitations to this study: (1) The sampling periods were limited to only two postdiversion observations and extended only 6 months. Six months is too short a period to determine the community effects of variables such as rainfall, seasonal effects, nutrient cycles, and biological cycles. (2) The settling plate studies did not have "prediversion" data, so a contrast between pre- and postdiversion settling cannot be made. No measurements of fish were performed prior to the discharge either. (3) Monitoring of a control site, where no salinity changes occurred, would have provided important baseline data. (4) Results of this study have limited applicability to other systems, but many of the biological communities are similar to those in the Tampa Bay area. There is about a 50 percent overlap in species and abundant seagrass communities at both locations, though there was only a 10 percent similarity of benthic infauna. Clearly, however, the study could not have been done in Tampa Bay itself. (5) The study included chemical analyses of metals, but not of other chemical additives in the discharged concentrate. The accumulation of these chemical additives in the sediments or organisms in the vicinity of the plume was not studied. Nevertheless, this study is the most comprehensive of its kind. Its limitations strongly suggest the need for standard rules and protocols for future studies.

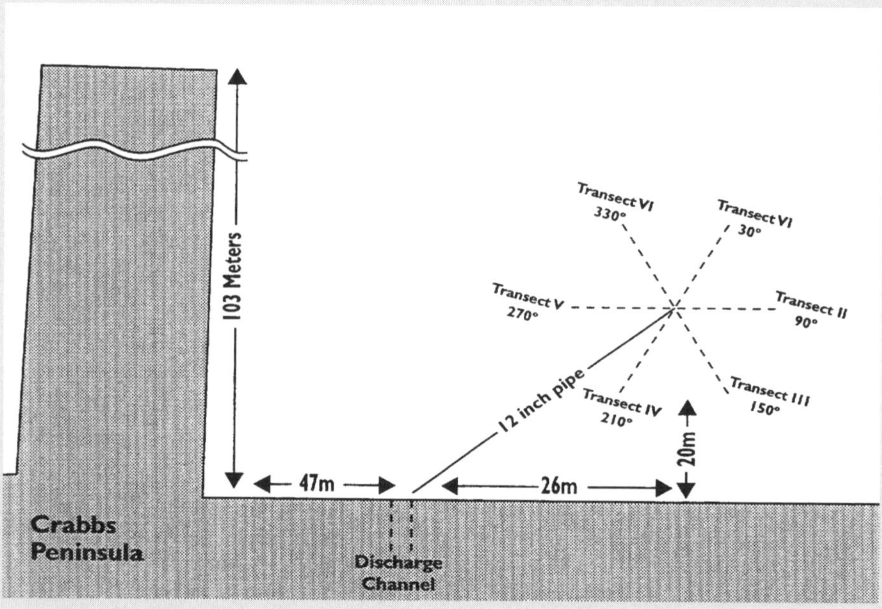

FIGURE 5-5. Diagram of sampling area for Antigua study.
SOURCE: Hammond et al. (1998).

Another study that provides additional useful information is an investigation of the ecological effects resulting from the application of granulated sea salt as a management technique to kill the invasive alga *Caulerpa taxifolia* (O'Neill et al., 2007). Although the salt treatment causes cell lysis and death of the target alga, nearby seagrasses (*Zostera*), their epiphytes, and nearby infauna showed no consistent effects on abundance or diversity.

From these studies, the effects of desalination concentrate discharge into oceans, seas, and estuaries appear to vary widely. The impacts depend on the site-specific environment, the organisms examined, the amount of dilution of the concentrate, and the use of diffuser technology.

Modeling and Monitoring Coastal Concentrate Discharge. Computer simulation models can be created and utilized to predict the environmental changes that may occur as a result of concentrate discharge (see Box 5-5). Through these predictive models, technical evidence can be produced to support applications for the NPDES permit and other environmental permits that may be required for concentrate discharge. Through this process, a utility can minimize environmental impacts from concentrate discharge and also generate greater public acceptance of the technology.

Additional environmental ecosystem monitoring is needed to better demonstrate the environmental sustainability of desalination using surface water concentrate discharge. Hydrobiological monitoring programs using scientifically accepted methods that allow utilities to track changes to coastal ecosystems over the operating period should be strongly considered. Pilot studies should be performed prior to construction of a plant, and baseline data should also be collected. After construction of the desalination facility, monitoring studies should be continued.

Protocols should be developed for monitoring programs using standardized procedures, including before-and-after studies, the length of transects to be examined, and the use of reference sites. A possible monitoring approach might be to follow transects perpendicular and parallel to the shoreline at both the impacted and reference sites, to survey the abundance, diversity, and health of resident planktonic, benthic, and nektonic (e.g., fish) species. Utilities should begin monitoring at least 1 year prior to the start of the projected operating period to establish a baseline of ecosystem data with which to compare postproduction data. Monitoring should be continued over a period of several years after the plant comes online and should include samples taken during different seasons of the year. When the receiving environment includes seagrass beds or kelp forests, they should also be carefully evaluated. Settling plates should be included for surveying attached organisms. Because the concentrate will tend to settle on the bottom, benthic community analysis (i.e., abundance, diversity, community structure) should be an important part of a monitoring program. There are well-established protocols and techniques for performing these analyses (Cao et al., 1996; Gaston and Young, 1992; Horne et al., 1999; Weis et al., 2004). The development of

BOX 5-5
Modeling Environmental Impacts of Estuarine and Marine Desalination

The regulatory approval and permitting process for seawater desalination requires the application of computer modeling to estimate the environmental effects of concentrate discharge. Various models have been developed for surface water concentrate and blending discharge and must be developed on a site-specific basis. This is necessary to provide an estimate of salinity changes within the three dimensions (depth and area) of the water body. Site-specific variation includes seasonal fresh surface water flows, tidal changes, water depth, and wind speed and direction. Once the modeled salinity changes are determined, an evaluation can be performed on the marine species to estimate health effects on those species. These models are referred to as three-dimensional (3D) time-dependent hydrodynamic models. Most of these models descended from the Princeton Ocean Model (Blumberg and Mellor, 1987). Comprehensive environmental monitoring programs are necessary to collect background data prior to the desalination operation and after the desalination project startup. Data from these monitoring programs can be used to calibrate the environmental models for future plant expansions or for other desalination projects in the vicinity.

One example of model application is the Tampa Bay Seawater Desalination project. In this case, a 3D hydrodynamic model for "far-field" effects was developed and applied based on an advanced version of the Blumberg-Mellor Estuarine Coastal Ocean Model (ECOM-3D) (Luther et al., 1998). Modeling for prediction of "near-field" effects was performed using MIKE 3, a 3D nonhydrostatic finite difference model developed and applied by the Danish Hydraulic Institute.

monitoring protocols is very important to facilitate environmental impact assessment studies. The development of a monitoring framework is also an objective of the European research project MEDINA (Membrane-Based Desalination, an Integrated Approach) funded by the European Commission.

Management of Inland Desalination Concentrates

Brackish desalination facilities that are a significant distance from the coast have unique considerations when it comes to concentrate management. Brackish desalination facilities operate at higher recovery efficiencies (50 to 90 percent) and therefore produce lower quantities of concentrate per equivalent volume of product water. Because the source water typically has significantly lower salinity than seawater, the concentrate from brackish water desalination plants also has significantly lower salinities than seawater desalination plants, and inland facilities that use brackish groundwater as source water require fewer pretreatment chemicals. Nevertheless, despite these apparent advantages, finding cost-effective approaches for concentrate management with minimal environmental impacts in inland locations remains a challenge. Alternatives to disposal are to treat the concentrates as potential economic products (e.g., selective salt recovery) that utilize the concentrates in some type of beneficial or commercial product (Ahmed et al., 2003; Jordahl, 2006). Environmental impacts of various approaches for inland concentrate management are discussed in the following section (see Chapter 4 for technical descriptions of the current concentrate management approaches).

Freshwater Rivers. Desalination concentrates are sometimes discharged to rivers and other inland surface water bodies in accordance with local, state, and national water quality regulations (Mickley, 2006). Additionally, 31 percent of municipal desalination facilities with greater than 95 m^3/day discharge directly to sewers upgradient of the wastewater treatment plant (Mickley, 2006). Most wastewater treatment plants ultimately discharge treated effluent to rivers, and while the traditional treatment process would remove many organic compounds, suspended solids, and some metals, it will not remove salts. However, the salinity in the concentrate will be diluted by the volume of treated wastewater. Inland communities that desalinate river water (or alluvial aquifers) and blend the concentrate with treated wastewater prior to river discharge should see only minor net downstream increases in river salinity caused by evaporative concentration and other salinization processes. However, concentrate from a deep brackish groundwater source could significantly alter the water quality of the surface water body. Depending on the source water composition, brackish groundwater concentrate may add toxic trace and radioactive constituents leached from the subsurface, such as selenium, arsenic, uranium, or radium. The potential for major ion toxicity, due to "imbalance" of ions in the concentrate, discussed previously, is also a concern when concentrate from brackish groundwater is discharged into freshwater ecosystems.

Impacts of discharge of brackish water to surface water sources would be expected to be greater in freshwater than in estuaries or the marine environment, where salt is a natural component of the ecosystem. Some freshwater organisms are only able to tolerate low levels of dissolved solids. If salinity increases in the water body, a shift to more salinity-tolerant species can be expected. High salinity may interfere with the growth of aquatic vegetation. Salt may decrease the osmotic pressure, causing water to flow out of the plant, resulting in stunted growth. Water containing high salt concentrations may create brackish layers in receiving lakes. Since saltwater is denser than freshwater, it tends to sink and form a layer at the bottom that does not mix with remainder of the lake water, leading to decreased dissolved oxygen levels.

Elevated salinity is stressful to many freshwater organisms. Sarma et al. (2005) studied freshwater crustaceans (anostracans) that inhabit ephemeral water bodies in which the water level decreases due to evaporation, increasing the salt concentration. They found that increased salinity resulted in decreased survivorship. Females showed several peaks of reproduction at 0 and 1 ppt salinity, whereas at 4 or 8 ppt there were fewer peaks. The highest reproductive rate was in 0 ppt of salt, while the lowest was at 8 ppt. Average lifespan, life expectancy, gross and net reproductive rates, generation time, and the rate of population increase were inversely related to the salt concentration. Freshwater species from environments that do not normally experience increased salinity would likely be much more susceptible than these anostracans.

To minimize environmental effects, concentrate discharge to rivers (and sewers) needs to be coordinated with background water quality, the composition of the concentrate, discharge rates, blending characteristics, and local water quality standards. Sewer discharge of concentrate upstream of the wastewater treatment plant should also be managed so as not to exceed the capacity of the treatment plant or to adversely impact its biological processes with excessive salinity.

Evaporation Ponds. Potential environmental impacts from the use of evaporation ponds include leakage of the concentrate and degradation of underlying aquifer systems or adjacent freshwater resources. Engineered low-permeability barriers are used to reduce the likelihood of leakage from the pond. Other factors that affect environmental water quality include sufficient basin storage volume to prevent overflow in case of major precipitation events, and location of sites topographically above long-term flood reoccurrence intervals of nearby water sources. The elevated salinity and trace constituents in evaporation ponds may be problematic for breeding and migrating birds, as was seen with the selenium effects on birds at the Kesterson National Wildlife Reserve (Hannam et al., 2003; Hoffman et al., 1988; NRC, 1989).

Land Application. As described in Chapter 4, the allowable salinity for land application depends on the tolerance of target vegetation, percolation rates, and the ability to meet the groundwater quality standards and is, therefore, more viable for lower salinity concentrate. In many cases, additional dilution water is needed for land application to be feasible. It may be possible to genetically engineer better salt-tolerant plants in the future and utilize these plants for animal fodder (Grattan et al., 2004; Grieve et al., 2004). However, if transpiration from the plants exceeds precipitation to the soil, over time any salts not taken up by the plants will accumulate in the soil. If the source water, and thus the concentrate, contains contaminants of concern such as arsenic, nitrate, or other harmful trace metals, the potential environmental impacts could include uptake of these contaminants by the plants or leaching of these contaminants into the soils or groundwater. Currently, in arid and semi-arid environments (generally west of the 100 Meridian in the United States), land application is not a physically sustainable method for disposal of desalination concentrate, because it is likely to exacerbate an already a large worldwide problem of soil salinization (NRC, 1993).

Injection Wells. Disposal of concentrates through injection wells is required to meet criteria established by the EPA for its Underground Injection Control Program (EPA, 2007b; see Box 5-1) to ensure that well injections do not endanger aquifers supplying drinking water by allowing the injected concentrate to enter the aquifer and degrade the resource. It should be understood that the amount of aquifer storage in a typical confined aquifer injection environment is small—about 1 m^3 per 10,000 m^3 of aquifer material—and thus, there will be displacement of current aquifer fluids. If disposal occurs in a depleted oil reservoir or an unconfined aquifer, storage would be much larger—about 1 m^3 per 10 m^3 of aquifer material. To prevent adverse impacts to surrounding aquifers, the volume, location, and solute composition of any displaced fluids and how they might influence the water quality of surrounding aquifers or surface waters should be well understood. This involves quantifying all flow boundaries and simulating groundwater flow dynamics using appropriate three-dimensional numerical transport and flow models (see Box 5-3).

Concentrate injection in artesian aquifer systems, which are typical of most formations used for deep-well injection, locally causes increase in fluid pressure and vertical expansion of the aquifer framework, which may be expressed as a rise in land surface. This increase in fluid pressure can also trigger earthquakes in certain geologic environments. Deep injection wells have caused several large-magnitude earthquakes (5 or greater on the Richter scale) and several thousand smaller ones in areas that are structurally stressed, such as the Rocky Mountains in Colorado (Evans, 1966; Hsieh and Bredehoeft, 1981) and Rangely oil field, Colorado (De la Cruz and Raleigh, 1972). Thus, proposed injection sites need to consider the potential for this condition if the target formation is deep and in an area that has experienced tectonic activity in the relatively recent geologic past.

Landfilling. One potential disposal option is to convert the concentrate from a liquid to a solid (or a dense slurry) and then dispose of the waste material in a suitable landfill (see Thermal Evaporation in Chapter 4). It requires a great deal of energy, however, to remove and recover the liquids from the concentrate and then to transport the wastes to a landfill, and these requirements may have significant financial, social, and environmental ramifications (see Greenhouse Gases in this chapter). Because most landfills eventually leak, there are also potential future environmental impacts to groundwater near the landfill.

WATER QUALITY ISSUES IN DESALINATED PRODUCT WATERS[4]

Because desalination processes employ advanced water treatment techniques, it is commonly assumed that desalinated water is devoid of contaminants. In reality, although desalination technologies remove various constituents to a large extent, not all constituents are fully removed and some species are removed to a lesser extent than others. In RO, a small fraction of ions, especially monovalent ions such as sodium and chloride, and dissolved organic molecules (e.g., some pesticides or herbicides) can pass through to the permeate water. Desalinated product water quality depends on the raw water quality, the treatment technology selected (e.g., RO, distillation, electrodialysis), and within membrane technologies, by the specific membranes employed and the implementation of second-pass RO. Boron and bromide are two inorganic constituents associated with water quality concerns in RO desalination, and these challenges along with approaches to mitigate these concerns are described next.

Boron occurs in the oceans at an average concentration of 4.5 mg/L (Weast et al., 1985). Although thermal desalination removes boron, the rejection of boron in RO desalination is dependent on the pH. Rejection increases with pH, although the single-pass RO process is operated at a low pH to avoid scaling. Single-pass RO desalination processes do not remove the majority of boron in the raw water at typical operating pH ranges; thus, boron (occurring as borate or boric acid) can be found at milligram-per-liter levels in the finished water. Implementation of a second pass through RO membranes with a pH adjustment to place boric acid in its negatively charged borate form can provide effective boron removal (Karry, 2006; Magara et al., 1998). Second-pass RO installation and operation, however, have significant cost implications and historically are not routinely included in desalination projects.

---

[4] Minor changes have been made to this section, a related conclusion at the end of the chapter, and an associated research recommendation in Chapter 8 after release of the prepublication version to incorporate data on boron toxicity and exposure levels from the 2000 IOM report, *Dietary Reference Intakes for Vitamin A, Vitamin K, Arsenic, Boron, Chromium, Copper, Iodine, Iron, Manganese, Molybdenum, Nickel, Silicon, Vanadium, and Zinc.*

Although boron is recognized to have a beneficial role in some physiological processes in some species, higher exposure levels may cause adverse health effects (IOM, 2000). No human health effect data are available on adverse health effects from ingestion of large amounts of boron from food and water, although data are available on the human health effects of large and small doses of boric acid or borax. However, most of the boron toxicity data come from studies in laboratory animals. High-dose boron exposures had the greatest effect on developing fetuses and on testes and led to reduced fertility in experimental animals (IOM, 2000). The EPA concluded that there is inadequate data to assess the human carcinogenicity of boron (EPA, 2006). Some research on the environmental effects of boron has shown that boron at milligram-per-liter levels also can adversely impact crops and grass species (Yermiyahu et al., 2007).

Boron was considered in the EPA's second Drinking Water Contaminant Candidate List (CCL2).[5] In the CCL2, the EPA defined a reference dose (or the level of lifetime exposure at which no adverse health effects are expected) of 0.2 mg/kg/day, conservatively estimated based on developmental effects in rats as well as applied uncertainty factors based on the extrapolation of data from animals to humans (EPA, 2006). Using the same toxicity data, but slightly different assumptions about uncertainty factors, the Institute of Medicine (2000) recommended a modestly higher exposure level of 0.32 mg/kg/day (or 20 mg/day for adults; 17 mg/day for adolescents, and exposures ranging from 3 to 11 mg/day for children ages 1-13).

Translating these exposure guidelines into recommended limits for boron concentrations in water requires assumptions regarding other possible sources of boron. According to IOM (2000), airborne boron contributes very little to the daily exposure of the general population. For humans not taking dietary supplements, diet is the major source of boron intake followed by drinking water. In the U.S., the median intake of dietary and supplemental boron was estimated to be approximately 1.0 to 1.5 mg/day for adults (IOM, 2000).[6] Based on its calculated boron reference dose and an assumption that 20 percent of total daily boron consumption would come from drinking water, the EPA developed a health reference level for drinking water as 1.4 mg/L boron. With different assumptions of the total amount of boron exposures from drinking

---

[5] See http://epa.gov/safewater/ccl/ccl2.html#chemical.

[6] Ninety-fifth percentile dietary intakes of boron in the U.S. are approximately 2.3 mg/day and 1.6 to 2.0 mg/day for men and women, respectively; 2.7 mg/day and 4.2 mg/day for vegetarian men and women, respectively (Rainey et al., 1999). The average intake of supplemental boron at the ninety-fifth percentile is approximately 0.4 mg/day for adults (IOM, 2000).

water and other sources, this water quality guidance can vary. The State of California has adopted a notification level for boron at 1 mg/L (California Department of Public Health, 2007). The current World Health Organization (WHO, 2004) guideline for boron in drinking water is 0.5 mg/L, but this is due to be reconsidered under the rolling revision of the guidelines (WHO, 2007). A recent draft WHO report notes that the revised health-based guideline (anticipated in 2008) might be 1 mg/L or higher (WHO, 2007).

Because boron is not likely to be found at levels of concern in surface waters and groundwater, the EPA also made a preliminary determination not to regulate boron with enforceable drinking water standards (see Box 5-1; EPA, 2007a). Instead, the EPA encouraged states with public water systems that have boron at concentrations higher than 1.4 mg/L to evaluate site-specific protective measures and to consider whether state-level guidance or regulation is appropriate.

Although boron is not specifically regulated in product water in the United States, consumer expectations may pressure desalination planners to design future seawater plants to follow these current guidelines. Treatment to these levels will increase the cost of new seawater desalination plants. Additional analysis of the human health effects of boron in drinking water, considering other sources of boron, are needed to support firm state-level water quality guidance for seawater desalination process design that is suitably protective of public health. If seawater desalination becomes a significant source for drinking water supply in the United States, additional regulatory attention or national guidance may be needed.

Because it is difficult for RO technologies to meet current boron guidelines in single-pass operations if there is boron in the feedwater, membranes and processes are being developed to reduce the level of boron in the product water. In some areas, specific resins combined with a small-scale RO are used to reduce the amount of boron. Boron can also be removed by optimization of RO, such as via multistep desalination or by coprecipitation with hydroxides (Cotruvo, 2005; Hyung and Kim, 2006). A technique for boron removal through reacting seawater with fly ash and coal materials has also been developed (Vengosh et al., 2004). Future seawater desalination projects should consider boron treatment options early in their planning efforts when considering the various end uses of the water produced.

Bromide is another water quality consideration for membrane desalination projects. Bromine is formed by the reaction between bromide and free chlorine, which is often used as a biocide to control biological growth in the intake and pretreatment systems for seawater desalination plants. Bromine in its uncharged form (HOBr) passes through RO membranes and is found in permeate water. Bromine participates in the formation of disinfection by-products (e.g., bromoform, dichloro-bromomethane, dibromochloromethane) when it reacts with natural organic matter (Laine et al., 1993; Singer, 1999; Summers et al., 1994). These by-products may have adverse human health effects (Richardson, et al., 2007) and are regulated through the SDWA (see Box 5-1) Disinfection By-product Rules. RO membranes have relatively low rejection capability for trihalomethane disinfection by-products; thus, some of these compounds pass through the membranes and reside in the permeate. Brominated by-products may also be formed if the desalination product water containing bromine is blended with water from other traditional sources containing natural organic matter. Bromide can also adversely affect the stability of chloramine in finished waters (Duirk and Valentine, 2007).

To minimize disinfectant by-product formation, chlorine is generally used only intermittently during pretreatment, but, in some cases of high organic loading, this may not be possible. At the Tampa Bay Seawater Desalination Plant, chlorine dioxide is utilized to control biological growth in the pretreatment process. The plant had previously used free chlorine but the disinfectant was changed due to elevated disinfection by-product formation. During the post-treatment process, monochloramines are formed as a secondary disinfectant to further reduce the disinfection by-product formation. Most utilities that have switched from a free chlorine residual to a monochloramine residual have done so primarily to reduce disinfection by-product formation to comply with current and future disinfection by-product regulations (Dyksen, 2007). During the formation of monochloramines, the ammonia can also combine with bromine to form bromamines (Bousher et al., 1989). Bromamines are not recognized by the EPA as an approved drinking water disinfectant.

In summary, boron, bromide, and disinfection by-products can affect product water quality. All are controllable through treatment optimization, but that treatment could aversely affect the cost of desalination.

GREENHOUSE GAS EMISSIONS

Water resource management currently uses significant amounts of electrical and natural gas energy to capture, treat, and transport water. The California Energy Commission (2005a) estimates that capture, transportation, and treatment of water uses approximately 5 percent of the electrical energy consumed in the state. Because of climate, geology, topography, and long water conveyance routes, the energy use for capture, transportation, and treatment in California is higher than the national average of 3.5 percent of electrical energy consumed (U.S. Department of Energy, 2006). Desalination is an energy-intensive process that would add more demand. A comparison of energy use for different water sources (Table 5-2) suggests that seawater RO requires about 10 times more energy than traditional treatment of surface water (Cohen et al., 2004). Concerns over anthropogenic climate change have spurred interest in the energy requirements of desalination. Although the percentage of statewide energy use is likely too small for desalination, planners will need a clear understanding of the energy and climate implications of desalination relative to other water supply alternatives as the nation takes steps to address the issue of greenhouse gas emissions.

Energy sources other than fossil hydrocarbons can provide energy for desalination and thus avoid or significantly reduce greenhouse gas emissions. Technologies such as nuclear (19.3 percent of electrical power in the United States), hydroelectric (6.5 percent), wind (<1 percent), and solar photovoltaic (<1 percent) are providing input to the electrical grid (Edison Electrical Institute, 2005) and are not associated with the generation of greenhouse gases. Other alternative energy sources such as biofuels (1.6 percent) are nearly neutral in terms of greenhouse gas emissions (Adler et al., 2007), and closed-loop geothermal systems can be essentially greenhouse gas-free. As discussed in

TABLE 5-2. Comparison of Energy Use for Different Water Sources in California.

Water Source	**Energy Used per Cubic Meter of Water (kWh/m^3)**
Pumping groundwater 120 ft	0.14
Pumping groundwater 200 ft	0.24
Treatment of surface water	0.36
Brackish water desalination	~0.3 to 1.4
Water recycling (no conveyance)	~0.3 to 1.0
Conveyance of water (examples):	
Colorado River Aqueduct to San Diego	1.6
San Francisco Bay Delta to San Diego	2.6
Seawater desalination (no conveyance)	~3.4 to 4.5

NOTE: Numbers reflect cited case-study examples and are not statewide averages.
SOURCE: Cohen et al. (2004), reprinted with permission from the National Resources Defense Council.

Chapter 4, thermal desalination plants can utilize low-grade or waste heat resources and substantially reduce their prime energy demands.

Commercial applications of alternative energy sources to power desalination remain somewhat limited. A 125,000 m^3/day membrane desalination facility in Perth, Australia (Water Corporation, 2007), that began operation in 2007 is the first example of using alternative energy to power desalination at a large scale. The Perth wind farm is not a dedicated stand-alone power source; rather it feeds into the power grid from which the desalination plant contracts to withdraw its electrical power. Waste heat from Japanese nuclear plants has been used to generate boiler water for the plants' own use, but no dedicated nuclear power plants have yet been developed for the purpose of powering water desalination (IAEA, 2005; Minato and Hirai, 2003; Pankratz, 2005).

A review of the potential for alternative energy for desalination (European Commission, 1998) and discussions of alternative energy for remote offgrid areas (García-Rodríguez, 2003; Tzen and Morris, 2003) suggest that several alternative energy sources hold promise. A variety of alternative energy sources have been proposed for various locations, depending on local conditions. These include photovoltaic (Richards and Schäfer, 2003) and heat-driven processes, such as direct solar evaporation (Trieb et al., 2003), closed geothermal (Bourounia et al., 1999; Karytsasa et al., 2004), ocean thermal energy conversion, and salinity-gradient solar ponds (Lu et al., 2001). Solar-powered desalination coupled with water reuse is a centerpiece of Masdar, an initiative in the United Arab Emirates to build the world's first carbon-neutral city. Proposed mechanical-driven alternative energies for desalination include wind power (Liu et al., 2002), wave power, tides, and hydrostatic head. Thus, there are numerous alternative energy technologies available, and these technologies may be able to provide the right quality of energy for desalination while reducing overall greenhouse gas emissions. More research, however, is needed to analyze the alternatives for coupling desalination with alternative energy sources in both inland and coastal areas.

Climate Change and Desalination

There seems to be no question that climate change will significantly impact the water resources sector and, as such, will indirectly impact desalinization. A rise in sea level over tens of years may have adverse impacts on coastal aquifers from increased seawater intrusion. Direct impacts of rising ocean levels may over the lifetime of the project have some minor effect on desalination structures built adjacent to coastlines because current sea-level rise is approximately 2 mm/year (United Nations Intergovernmental Panel on Climate Change, 2007). Furthermore, storms associated with climate warming may be of either higher frequency or higher intensity. Depending on the location of the intake, the temperature of the water may increase slightly, requiring small changes to the desalination process. Although these direct impacts to desalination structures and processes appear to be small, they should be clearly understood prior to the design of a major desalination facility.

CONCLUSIONS AND RECOMMENDATIONS

Knowledge of the potential environmental impacts of desalination processes is essential to water supply planners when considering desalination among many water supply alternatives. All components of the water-use cycle should be considered, including source water impacts, the likely greenhouse gas emissions from the energy requirements of the desalination process, potential impacts from concentrate management approaches, and environmental health considerations in the product water. Ideally, these considerations should be compared against equally rigorous environmental impact analyses of water supply alternatives. The role of science and engineering is to clearly articulate the environmental impacts in a transparent manner so that society can make an informed decision after comparing the full economic costs—including environmental costs—and benefits among the various water supply alternatives (as discussed in Chapter 6).

Because of the limited amount of long-term research, there is presently a considerable amount of uncertainty about the environmental impacts of desalination and, consequently, concern over its potential effects. A variety of environmental impacts are possible with desalination. Seawater desalination can cause impingement and entrainment of marine organisms and create ecological impacts from concentrate discharge. Desalination of inland brackish groundwater sources could lead to groundwater mining and subsidence, and improper concentrate management practices can negatively affect drinking water aquifers and freshwater biota. Site-specific information necessary to make detailed environmental conclusions on the ecological impacts of both source water withdrawal and concentrate management associated with desalination is lacking. The limited studies to date suggest that the environmental impacts *may* be less detrimental than many other types of water supply, but definitive conclusions cannot be made until more research is done.

Site-specific assessments of the impacts of source water withdrawals and concentrate management should be conducted and the results synthesized in a national assessment of potential impacts. Adequate understanding of impacts of source water withdrawals and concentrate management results only from site-specific assessments. The ecological effects of concentrate discharge into the ocean appear to vary widely and depend on the site-specific environment, the organisms examined, the amount of dilution of the concentrate, and the use of diffuser technology. The ecological impacts of surface water intakes (i.e., impingement and entrainment) have been well studied for power plants but not for desalination plants, and these impacts will likely vary from place to place. General information on potential impacts from groundwater withdrawal and injection are available from decades of hydrogeologic studies for other purposes, but site-specific analyses are necessary to understand the impacts from a proposed facility. Once a number of rigorous site-specific studies are conducted, this information should be synthesized to develop an overarching assessment of the possible range of impacts from both seawater and brackish water desalination in the United States. A characterization of the volume, hydraulic properties, flow boundary conditions, and solute chemistry of the nation's brackish groundwater resources and a characterization of the spatial distribution, thickness, and hydraulic properties of aquifer systems suitable for concentrate injection, relying heavily upon existing data, would assist the financial and environmental planning process for inland desalination facilities.

Longer-term, laboratory-based assays of the sublethal effects of concentrate discharge should be conducted. Except for a few short-term lethality studies that do not give insight into long-term effects, research on the impacts of concentrate discharges on organisms in receiving waters has been minimal. Longer-term laboratory-based biological assays, running from weeks to months in duration, should evaluate impacts of concentrate on development, growth, and reproduction using a variety of different organisms, including those native to areas where desalination plants are proposed. These results should be put into a risk assessment framework.

Monitoring and assessment protocols should be developed for evaluating the potential ecological impacts of surface water concentrate discharge. Adequate site-specific studies on potential biological and ecological effects are necessary prior to the development of desalination facilities because biological communities in different geographic areas will have differential sensitivity. For large desalination facilities, environmental data should be collected for at least 1 year in the area of the proposed facility before a desalination plant comes online so that sufficient baseline data on the ecosystem are available with which to compare postoperating conditions. Once a plant is in operation, monitoring of the ecological communities (especially the benthic community) receiving the concentrate should be performed periodically and compared to reference sites.

Water quality guidance, based on an analysis of the human health effects of boron in drinking water and considering other sources of exposure, is needed to support decisions for desalination process design. There are concerns about boron in product water from seawater desalination because the boron levels after single-pass RO commonly exceed current WHO health guidelines and the EPA health reference level. A range of water quality levels (0.5 to 1.4 mg/L) have been proposed as protective of public health based on different assumptions in the calculations. The EPA has decided not to develop an MCL or health-based MCLG for boron because of its lack of occurrence in most groundwater and surface water and has encouraged affected states to issue guidance or regulations as appropriate. Therefore, most U.S. utilities lack clear guidance on what boron levels in drinking water are suitably protective of public health. Boron can be removed through treatment optimization, but that treatment could aversely affect the cost of seawater desalination.

Further research and applications of technology should be carried out on how to mitigate environmental impacts of desalination and reduce potential risks relative to other water supply alternatives. For example, intake and outfall structures could be designed to minimize impingement and entrainment and encourage improved dispersion of the concentrate in coastal discharges. Research could also explore beneficial reuse of the desalination by-products and develop technologies that reduce the volume of this discharge. There are numerous alternative energy technologies available, and these technologies may be able to provide the right quality of energy for desalination while reducing overall greenhouse gas emissions; however, research is needed to analyze the alternatives for coupling desalination with alternative energy sources in both inland and coastal areas. Additional research investments should be able to clarify the potential risks of desalination and develop approaches to substantially mitigate the environmental impacts. Nevertheless, desalination efforts do not need to be halted until this research is done and uncertainties removed.

6

The Costs and Benefits of Desalination

The promise of desalination to rid the world of water scarcity has been touted for nearly 50 years. During this period, public and private investment in developing and improving desalination technology has totaled more than a billion dollars (see Chapter 2). Although much progress has been made and there have been successes in developing water supplies in very dry locales and regions, the promise remains largely unfulfilled. The explanation lies with the fact that, although the process costs have been reduced, the total costs of desalination, including the costs of planning, permitting, and concentrate management, remain relatively high, both in absolute terms and in comparison with the costs of other alternatives.

In assessing the future prospects and promise of desalination technology, it is particularly important to examine the current and prospective financial and economic circumstances that are likely to surround the technology as it develops. An examination of the structure of desalination costs and of the determinants of those costs is important in identifying areas in which research might be pursued with the greatest effect. A consideration of the availability and costs of alternative supplies helps to place the future role of desalination in perspective. Finally, issues of reliability, water quality, and environmental impacts need to be understood if the costs and benefits of desalination technology are to be broadly understood. All of these topics are considered in this chapter.

GENERAL CONSIDERATIONS

In assessing the economic desirability of any water supply project or facility, several general considerations and economic principles need to be understood. These principles, including the relative nature of cost-benefit comparisons, financial versus economic costs, and public versus private costs and benefits, are discussed below.

Relative and Absolute Costs

When evaluating water supply alternatives, it is important to understand that the benefits and costs should be evaluated in a relative rather than an absolute fashion. When the focus is on costs, for example, the absolute cost of the facility or project in question has little meaning unless it is compared with the costs of other alternatives for accomplishing the same purposes, if such alternatives exist. Thus, from an economic standpoint, employment of any desalination technology to augment water supplies will be attractive in circumstances where the alternatives are more costly or nonexistent, and unattractive in circumstances where less costly alternatives are available.

Economic and Financial Costs

In any assessment of economic viability, it is important to understand that economic costs and benefits will often differ from financial costs and benefits. Rarely, if ever, will a financial analysis be the same as an economic analysis. Economic analyses, including the analyses of costs and benefits, account for all of the costs to whomever they may accrue, irrespective of whether these costs are characterized by market-generated prices. Financial accounting and, consequently, financial feasibility are directly concerned with the availability and the costs of funds. Thus, for example, a utility or private-sector entity may be able to attract capital at a reasonable interest rate for the purposes of constructing a project if the returns will be sufficient to pay the interest charges, repay the principal of the loan, and allow appropriate return to management, labor, and other factors of production. In these circumstances, the project would be said to be financially feasible. There may, nevertheless, be costs or benefits that cannot (and need not, in this case) be captured because they are not internalized in the utility's cost stream or, in the case of benefits, not priced and subject to market-like exchanges.

Consider a situation in which there are large environmental costs of a water supply project that do not have to be addressed by law or regulation. Because these costs do not have to be addressed, they do not change the financial attractiveness of the project, but the environmental impacts are costs nonetheless, which must be borne by some individual or group. Hence, they must be counted as economic costs even though they do not affect the financial feasibility. Similarly, if some of the benefits cannot be captured by the project's operator, those benefits are economic benefits and must be counted as such in an economic analysis even though they will have no effect on the financial feasibility of the project.

Public and Private Costs

It is also important to draw the distinction between private costs and public costs. Private costs are internalized within the operation of the project and must be borne by the utility or business operator. They include costs such as wages, interest payments, energy, and equipment. Public costs, by contrast, are real costs that a business operator can escape but that will have to be borne by the public at large. In the absence of regulations, business operators usually have no incentive to mitigate or defray the costs of environmental damage that may accrue offsite as a consequence of their operations. In these circumstances, the costs will be borne by the public at large and are correctly accounted for as public costs.

A similar distinction should be made on the benefit side—the distinction between public benefits and private benefits. The benefits of a project are private benefits to the extent that they can be captured by the producer. When the producer sells the product, private benefits are captured through the price of the sale and are fully appropriable by the producer. By contrast, when benefits are jointly conferred on consumers and cannot be fully captured by the producer, they are said to be public benefits. Stated differently, if a producer is unable to withhold the benefits of the good or service from a consumer because the consumer refuses to pay for it, the benefits are public in nature. Such benefits tend to be undersupplied by producers, from a strict efficiency perspective, because they cannot capture those benefits fully. Environmental services (e.g., the capacity to purify water and air, to provide environmental stability, to protect against disease) are examples of public benefits.

Joint Costs

Finally, it is important to understand the notion of joint costs and the problems they create for the accounting and allocation of costs. Frequently the financial and economic costs of desalination can be reduced by combining desalination operations with the production of some other water-related goods and services. Power is the primary example; these dual-purpose plants are termed cogeneration facilities. Some elements of the plant will jointly serve both the production of power and the production of water. The costs of these elements that contribute to both purposes are defined as joint costs. In theoretical terms, it is not possible to allocate joint costs back to individual purposes in any but an arbitrary way. That is, it is not possible to partition joint costs according to the contribution to the marginal product of each of the multiple purposes. There are principles and informal rules and practices for allocating joint costs, including the separable cost/remaining benefit method, methods based on energy production and consumption, and others, but these are all arbitrary from an economic standpoint (Friedman and Moulin, 1999). Thus, while there may be controversy associated with allocating a disproportionate share of the costs to one product (such as power) rather than another (such as water), there is no theoretical principle or analytical reason which bars it.

The problem of allocating joint costs makes it particularly hard to analyze the costs of thermal desalination plants in a consistent and systematic way. Many thermal desalination facilities are co-located with power plants so that the waste heat from the power plant can be utilized in the desalination process, and others are designed as cogeneration facilities with large components of joint costs. The prevalence of joint costs in thermal facilities confounds efforts to allocate costs among the different inputs and to analyze the sensitivity of total costs to changes in various components of cost. For this reason, thermal technologies are largely omitted from consideration in the following section on the structure of costs.

A fully adequate economic assessment of a water supply project or strategic plan would be conditioned by and account for the considerations identified earlier. Benefit-cost analyses are relative, and the results will vary from situation to situation. Both economic analyses and financial analyses have important roles in any project analyses, but one should not be confused with the other. Where public costs and benefits predominate, there may be a case for governmental involvement. Where benefits and costs are largely private, the case for governmental involvement will be far weaker (Cornes and Sandler, 1996; Oakland, 1987). Although each project will have its own particular characteristics, it is possible to identify in some detail the elements and structure of costs and benefits. In the next section, the structure of costs for desalination facilities is considered and enumerated.

THE STRUCTURE OF DESALINATION COSTS

A summary of what is known about the current status and trends in desalination costs are presented in this section, followed by a detailed analysis of the determinants of those costs. An analysis of the structure of desalination costs is important to identifying research areas with the greatest potential for reducing costs.

Difficulties of Estimating and Comparing Costs

The cost to treat seawater or brackish waters to produce potable water is a function of numerous variables, and the components of these costs are frequently difficult to ascertain precisely from the literature. Although selling prices are reported for many international public-private projects, data on the components of the total cost and price are not reported and not available because they are regarded as confidential information by firms in the business and because of regulatory and public policies. The confidential nature of this information reflects the competitive nature of the international water business. Although water rates (or tariffs) are public information, these rates reflect the project-specific evaluation criteria, scope of work, and technical process impacts based on local conditions and requirements; therefore, they are not consistent from country to country or place to place. Consequently, tariffs do not provide a simple indicator for cost comparisons.

Different project costs are also difficult to compare because virtually every desalination plant has its own unique design and site conditions and its own unique financing package. Table 6-1 provides an example of such comparative costs for three projects: the desalination facilities built and operated by the Inland Empire Water Agency in southern California for the purpose of desalting brackish water; the brackish water desalination project in Texas developed by the El Paso Water Utilities in cooperation with the U.S. Army (see Box 5-2); and the Tampa Bay seawater desalination plant in Florida. Although it is tempting to draw conclusions from comparisons such as these, particularly with respect to the sensitivity of costs to source water salinity, great care must be exercised. For example, sometimes there are financing offsets that lower the apparent costs to the end users. Both Inland Empire and Tampa Bay will receive such offsets (\$0.20/m^3 or \$250/acre-foot [a.f.] to Inland Empire from the Metropolitan Water District of Southern California; \$0.09/m^3 or \$111/a.f. to Tampa Bay from the Southwest Florida Water Management District), although these offsets are not factored into the costs reported in Table 6-1.

Reviews of published data on costs can be confusing because costs are rarely reported consistently and some cost parameters are not reported at all. Additionally, the underlying assumptions (e.g., project life, project size) may differ and sometimes remain unstated (Almulla, 2002; Busch and Mickols, 2004; Dreizen, 2006; Frenkel, 2004; Hinkebein and Price, 2005; Miller, 2003). For example, some cost data include distribution costs while others are for costs at the plant boundary. Miller (2003)

TABLE 6-1. Financial Costs from Three Desalination Facilities

	Inland Empire	El Paso/Ft. Bliss	Tampa Bay
Feedwater total dissolved solids (TDS) (ppm)	800-1,000	1,200-15,000	26,000
Average output (m^3/d)	27,000	100,000	95,000
Operations and maintenance (\$/m^3)	0.31		0.45
Admin and general (\$/m^3)	0.029		0.02
Capital consumption (\$/m^3)	0.19		0.22
Fixed costs (\$/m^3)	0.10		0.15
Total (\$/m^3)	0.63	0.43	0.83

NOTE: The figure for the El Paso/Ft. Bliss project includes distribution costs whereas the figure for Tampa Bay does not.
SOURCE: R. Atwater, Inland Empire Water Agency, personal communication, 2007; E. Archuleta, El Paso Water Utilities, personal communication, 2006; J. Maxwell, Tampa Bay Water, personal communication, 2007.

summarizes costs reported in the published literature for a variety of desalination projects and notes that the numbers can only be used as a rough guide because they are not calculated on a consistent basis. Despite the limitations of proprietary data, there is a wealth of information available on the nature of desalination costs and on the ways in which those costs are determined. If this desalination cost information were compiled on a reasonably consistent basis, it would be particularly important to water planners who are concerned with problems of meeting growing water demands in the future. However, there is so much variation in the circumstances of individual projects, as well as in the bases upon which reported costs have been calculated, that the resulting numbers must be interpreted with great care and strict comparisons are not usually possible.

The information and analysis that follow result from efforts to provide a detailed view of the various cost components that make up the costs of producing freshwater from seawater and brackish water sources using membrane technology. The conclusions that can be drawn from the analysis are limited by the lack of consistency and detail in the data. The conclusions are also limited by the committee's inability to analyze the costs of concentrate management, including environmental costs, in anything but a general way. As noted in Chapter 5, the environmental costs of concentrate discharge are not well understood, and available concentrate management alternatives, and their associated components of costs, vary greatly from situation to situation. In this report, desalination costs that do not include the costs of concentrate management are specified as desalination production or process costs. It is also important to acknowledge that these challenges make it virtually impossible to generalize in any meaningful way about the structure of costs of inland facilities that utilize brackish feedwaters. In general, it is known that energy use and costs vary directly as a function of the concentration of total dissolved solids (TDS) of the feedwater (see Figure 4-10). Thus, where other things are equal, brackish waters should be less costly to desalinate than seawater. The problem is that brackish water desalination costs vary significantly, not only with the TDS of the feedwater but also with the costs of concentrate management, which can be very high in inland situations where brackish water desalination is otherwise attractive. For these reasons, brackish waters are not treated with the same level of detail or rigor in the cost analyses that follow.

Reported Desalination Costs

In general, the unit costs of producing freshwater from seawater have been reported in a range running upward from approximately $0.64/m^3 ($800/a.f. or $2.46/thousand gallons [kgal]) (see, for example, Miller, 2003). Many of these estimates, particularly those at the lower end of the range, include subsidies or do not account fully for all costs (Miller, 2003). Some large seawater desalination plants appear to be operating in the range of approx $0.80/m^3 ($3.06/kgal; $1000/a.f.) but new facilities are being proposed with substantially higher costs due to site-specific considerations (GWI, 2006a; Miller, 2003). The only operational examples where substantially lower production costs have been achieved entail the use of membrane technologies and brackish feedwaters with TDS concentrations that are significantly lower than that of seawater. Treatment systems for such lower-salinity groundwaters can often produce water for less than half the costs of treating seawater. However, when low-cost concentrate management methods are not available, brackish groundwater desalination costs can reach or exceed seawater desalination costs. It is important to note that desalination costs found elsewhere in the world may be lower than those that can be realized in the United States because the permitting costs and the costs of meeting environmental regulations tend to be high in the United States.

Membrane and thermal processes are both used widely in municipal-scale desalination plants worldwide and, as discussed in Chapter 4, each technology has strengths and weaknesses and differing operating conditions under which one or the other may be economically optimal. Thermal desalination systems consume more energy than reverse osmosis (RO) systems and are more capital intensive. Nevertheless, thermal systems can use more diffuse or low-grade forms of energy (i.e., low-pressure steam) whereas membrane systems rely solely on electricity as an energy source. Global Water Intelligence (GWI, 2006a) reports the capital costs of seawater desalination by multi-effect distillation (MED) and multistage flash (MSF) distillation to be 1.5 to 2.0 times the capital costs of RO desalination systems, respectively. Additionally, GWI (2006a) approximates the costs for the seawater desalination process by RO to be $\$0.61/m^3$ as compared to $\$0.72/m^3$ for MED and $\$0.89/m^3$ for MSF. The breakdown for these costs is shown in Table 6-2. These costs are based on a system scale of 100,000 m^3/day; a nominal interest rate of 6 percent; \$450 element cost; \$0.05/kWh energy cost; assumed electricity use of 4.5, 4.0, and 1.25 kWh/m^3 for RO, MSF, and MED, respectively; and a 20-year capital-payback period. The costs for seawater desalination by RO are slightly lower than costs reported from actual installations. Most likely this is due to the favorable interest and energy costs used in the preceding analysis. Additionally, the calculated total costs for thermal technologies are likely exaggerated because offpeak electricity costs, cogeneration, or the use of low-grade or waste energy are not considered in this analysis. If low-cost, dispersed sources of energy are available or energy can be jointly used with other purposes, seawater

TABLE 6-2. Comparative Total Cost Data for the Desalination Process for 100,000 m^3 of Seawater by Reverse Osmosis, Multistage Flash Distillation, and Multi-Effect Distillation

	SW RO	SW MSF	SW MED
Annualized capital costs	0.15	0.29	0.22
Parts/maintenance	0.03	0.01	0.01
Chemicals	0.07	0.05	0.08
Labor	0.10	0.08	0.08
Membranes (life not specified)	0.03	0.00	0.00
Thermal energy	0.00	0.27[a]	0.27[a]
Electrical energy (\$0.05 k/Wh)	0.23	0.19	0.06
Total ($\$/m^3$)	0.61	0.89	0.72

[a] The costs of thermal energy are likely exaggerated because offpeak electricity costs, cogeneration, or the use of waste energy are not considered in this analysis.
SOURCE: GWI (2006a).

desalination using thermal technologies becomes more cost effective. Leaving aside situations in which energy can be obtained cheaply, the capital and operating costs of thermal systems appear significantly higher than the best-available membrane technology.

As previously discussed, concentrate management costs can be widely ranging, based on the alternatives available, the volume and salinity of the concentrate, and other site-specific factors (see Chapter 4). There are a number of examples which illustrate the potential magnitude of concentrate management costs. One study, in which a zero liquid discharge (ZLD) option was evaluated as part of a desalination treatment train for an inland municipal application, estimated the cost of the ZLD steps as almost twice as large as the cost of the primary desalting step. The total desalination cost was estimated at \$1.44/m^3, of which the water production step accounted for \$0.50/m^3 while concentrate management costs amounted to \$0.94/m^3 (Sethi et al., 2007). For another project, currently under construction and scheduled to be online in 2008, the bid construction cost for a 3,000 m^3/day (0.8 million gallons per day [MGD]) RO facility was about \$26 million, of which approximately \$7 million was accounted for by the costs of a brine concentrator (Yallaly et al., 2007). Similarly, the brackish water desalting facility near El Paso, Texas, entailed significant costs for concentrate management. The projected cost of the facility was \$0.44/m^3 (\$1.64/kgal) of blended water including amortized capital, operation, and maintenance costs (assuming an energy costs of \$0.07/kWh). Of the capital cost of \$87 million, 26 percent was allocated for the concentrate disposal wells and lines. The estimated annual operating and maintenance costs for the concentrate management were lower, representing 0.04 percent of the estimated \$4.8 million costs (E. Archuleta, El Paso Water Utilities, personal communication, 2006).

Comparing Desalination Costs against Other Alternatives

When making water policy decisions, desalination costs need to be compared with the costs of other water supply or demand options available in a given locale. While there is some prospect that the costs of producing freshwater from seawater may come down, the existing evidence suggests that they are still quite high when compared with the costs of alternatives in most locales (see Figure 6-1 and Box 6-1).

Specifically, there are many instances in which demand management measures can make water available for new uses at costs that are significantly lower than the costs of desalination (See Box 6-1). Market-like transfers of water, in which water is reallocated away from relatively low-valued uses to relatively high-valued new uses, can also be less

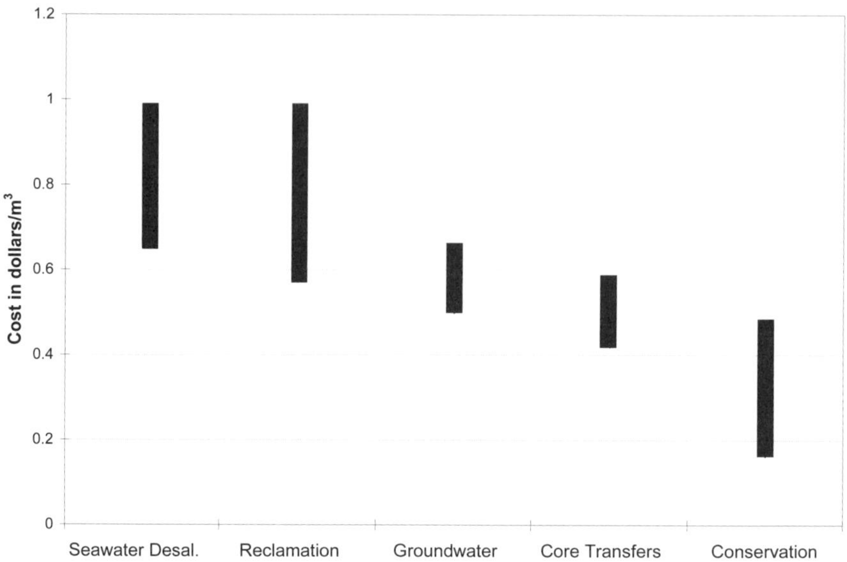

FIGURE 6-1. Financial cost ranges for a subset of available water alternatives for San Diego Water Authority in dollars per cubic meter, based on data from Robert Yamada, San Diego Water Authority, personal communication, 2006. These costs are *not* assumed to be representative of the nation as a whole but are provided as an example of a cost comparison among alternatives for one community. The high range for reclamation accounts for the additional costs of constructing more substantial conveyance facilities.

expensive than desalting in many instances. Although these methods result in the reallocation of some water among uses while desalination creates "new" water, the end result of the reallocating alternatives is a more efficient pattern of water use in which water itself is more productive than it was prior to the reallocation (Colby, 1997; NRC, 1992).

Comprehensive cost comparisons are difficult to make, because of uncertainty, rapidly changing technologies and prices, noncomparable variables, and other factors. Recent reviews (Chaudry, 2007; Cooley et al., 2006) suggest that, although energy costs have remained relatively stable over time (see Box 6-2), in the past 2 years energy price increases have begun to outpace desalination cost reductions due to improvements in technology. The problem is complicated not only by virtue of the fact that the total costs of alternatives vary with site-specific conditions but also by the fact that the total costs of desalination facilities differ from situation to situation. Factors such as the size of the plant, the salinity of the feedwater, the temperature of the feedwater, the prevailing costs of energy in the region, land costs, and the investment and operating costs

BOX 6-1
Costs of Conservation

Economists use a variety of methods to compare the relative costs of water supply and management alternatives, including cost-effectiveness, marginal costs, rates of return, and more. These different methods often produce different results depending on the point of view or perception of the target audience, the data used, and the assumptions made. For example, the cost of a conservation measure may differ substantially from the points of view of the water user as opposed to the water supplier. In addition, the economic benefits may also differ substantially from different perspectives, even assuming that all benefits can be put in comparable terms.

A typical approach is to use cost-effectiveness analysis to compare a unit cost of alternatives, for example in dollars per cubic meter of physical water supply or conservation reduction. A water conservation alternative is considered cost-effective when the unit cost of conservation (sometimes called "the cost of conserved water") is less than the unit cost of the next unit of additional supply.

A challenge in determining the costs of conserved water is the difficulty of identifying and quantifying many different cost factors. For example, the cost of conserved water depends on the capital cost and lifetime of the conservation technology (a low-flow toilet or front-loading washing machine, for example), but it also depends on the amount and cost of energy savings that might result, the amount and cost of changes in wastewater treatment costs resulting from a decrease in overall water used, and the value of restored ecosystem water flows, if any, among other factors. Replacing lawn with xeriscaping may save water, and it may also save labor and energy costs associated with reduced frequency of mowing grass. Properly identifying and calculating all of the associated benefits of water conservation options is important in order to properly compare among alternative water choices.

It is also important to distinguish between natural and accelerated replacement of water-using options. Natural replacement refers to the replacement of water devices due to age or failure; accelerated replacement refers to replacement of a device before the end of its natural lifetime specifically in order to reduce water use.

Perspective is important. For water conservation analyses, it has been argued that the proper perspective is the viewpoint of the water consumer, as opposed to the more traditional perspective of the water supplier (see, for example, Chapter 5 in Gleick et al., 2003). Analyzing the cost-effectiveness from the perspective of the consumer requires calculating the cost of conserved water based on the investment required of the consumer and any changes in operation and maintenance costs that the consumer would experience from the investment. These costs must then be compared to the marginal costs of alternatives. This approach fairly compares the costs and benefits to the water supplier (which are passed on to the consumer) with the costs and benefits experienced by customers apart from what they pay for water services alone, which often fail to account for associated benefits such as energy, labor, or other savings.

When the cost of conserved water from a specific measure is less than the cost of water supply displaced by conservation, the customer and the water utility (collectively) will make money if the measure is implemented (Gleick et al., 2003). If water rates and utility rebates do not accurately and fully reflect the

continued

marginal costs of supply, the net benefits to consumers and the utility may not be apparent. This can be corrected by adjusting water rates. It is thus important to properly identify both the marginal variable costs and marginal capital costs of water supply displaced by efficiency improvements.

A comprehensive analysis of water conservation alternatives was produced in 2003 by the Pacific Institute for California residential, commercial, and industrial efficiency options. The results of that study were presented as "supply curves for conserved water" in a form similar to that used in the field of energy economics (CPUC, 2001; Koomey et al., 1995). They concluded that as much as 2.5 billion m^3 of water could be conserved for less than $0.50/$m^3$ of water—a benchmark chosen because it is a commonly used measure for the marginal cost of new water supply and because no significant new water project in California has been proposed below that cost. They note, however, that at least 810 million m^3 of conservation is cost-effective to conserve in the residential sector even if new supplies cost only $0.05/$m^3$ (Gleick et al., 2003).

How is this low cost possible? It turns out that the estimated costs of some conserved waters are actually negative. This means that water could be free (have zero cost) and consumers would still save money by implementing the conservation option. For these options, nonwater benefits (such as savings in labor, energy, wastewater treatment, or fertilizer or pesticide use) are sufficient by themselves to pay for the efficiency improvements. For example, the cost of replacing inefficient toilets in restaurants and supermarkets was estimated to be on the order of $0.08/$m^3$, but the cost of replacing low-flow showerheads in hotel rooms was estimated to be $-0.65/$m^3$ because of the substantial energy savings associated with the reduction in hot water use. Similarly, improving housekeeping practices and using a dry-clean-up system in meat processing plants is not cost-effective based on the water savings alone ($1.10/$m^3$), but becomes highly cost-effective ($-0.50/$m^3$) when savings in wastewater treatment costs from biological oxygen demand fees are included. Water-efficient dishwashers in restaurants become tremendously cost-effective when the costs of energy and chemicals saved are properly included (Gleick et al., 2003; see, especially, Table 5-15).

of particular environmental mitigation facilities all combine to make the total cost of each facility different.

Examining the Determinants of Costs of Reverse Osmosis Desalination

Given the difficulties in making broad conclusions based on reported desalination costs, the committee took an alternative approach and conducted an original analysis that illuminates the key determinants of desalination cost and examines the sensitivity of cost to variations in these determinants. This analysis was conducted to illuminate the best opportunities for additional cost reductions, and thereby to serve as a framework for focusing future research and development efforts aimed at lowering the costs of desalination through technology innovations.

Sensitivity Analysis Methods

The following analyses are based on a set of baseline capital and operating cost data that were obtained for actual operating RO desalination systems of three different scales (38,000 m^3/d, 189,000 m^3/d, and 380,000 m^3/d) (M. Shah, GE Global Research, personal communication, 2007). Energy costs and membrane life in these baseline data were $0.07/kWh and 5 years, respectively. Process recovery was set at 40 percent, and the operating pressure was set at an average of 5,516 kPa (800

BOX 6-2
The Importance of the Price of Energy

The costs of desalination have always been sensitive to the price of energy because desalination processes, whether thermal or membrane, are energy intensive. Although it is possible that the energy intensity of membrane processes can be reduced, as discussed in this report, the likelihood is that the price of energy will continue to play a major role in the costs of desalination for the foreseeable future. The costs of electricity for the 46-year period 1960-2006 are shown in Figure 6-2. While it is true that electricity costs are not always the pertinent costs for thermal technologies, the trends in these costs are generally representative of energy costs for thermal technologies and reasonably exact costs for membrane technologies. The data in Figure 6-2 reveal several important conclusions.

First, and somewhat surprisingly, the costs of electricity in the period 2004-2005 were not higher than the costs in 1960 in constant dollar terms. This means that while there may have been ups and downs, the general electricity price picture for the period was relatively favorable for electricity users. Second, although there was a trough in the early 1970s and a peak in the early 1980s, electricity costs have been trending down since about 1985. During the intervening period of about 20 years, the costs of electricity became increasingly favorable in terms of their impact on desalination technology. Third, recently the costs of electricity have begun to trend upward. The price of oil on international markets has risen, and this suggests that there is reason to expect that electricity costs will continue to rise for the indefinite future. Although electricity costs are not directly tied to the costs of oil, because considerable quantities are generated from hydroelectric and coal-fired power plants, oil prices will provide a rough approximation of what is likely to happen to the cost of electricity. A rising trend in electricity prices is likely to affect the costs of desalination adversely unless there are parallel offsetting reductions in other costs. As noted in this chapter such reductions are quite possible.

The costs of energy, adjusted for inflation, are presented in Figures 6-2 and 6-3. These data show that real oil prices peaked in the late 1970s and early 1980s and then were relatively stable until 2002, when they began to rise markedly. Similarly, Figure 6-2 shows that electricity prices, which are typically less volatile than world oil prices, began to rise around 2002. If these price increases continue, they will slow the rate at which desalination costs can decline due to technological improvements.

continued

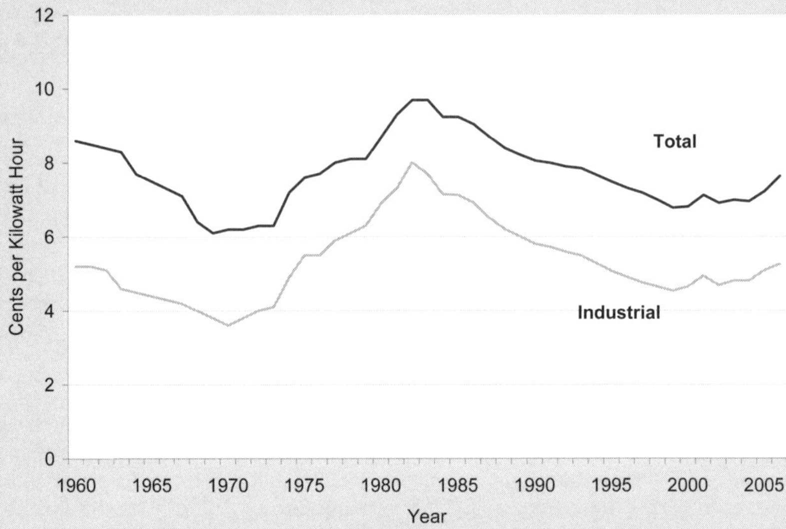

FIGURE 6-2. Inflation-adjusted electricity costs in the United States from 1960 to 2006 for all sectors and for the industrial sector alone. Prices are in cents per kilowatt-hours in constant 2000 dollars and include relevant taxes.
SOURCE: Data from U.S. Energy Information Administration (2007a).

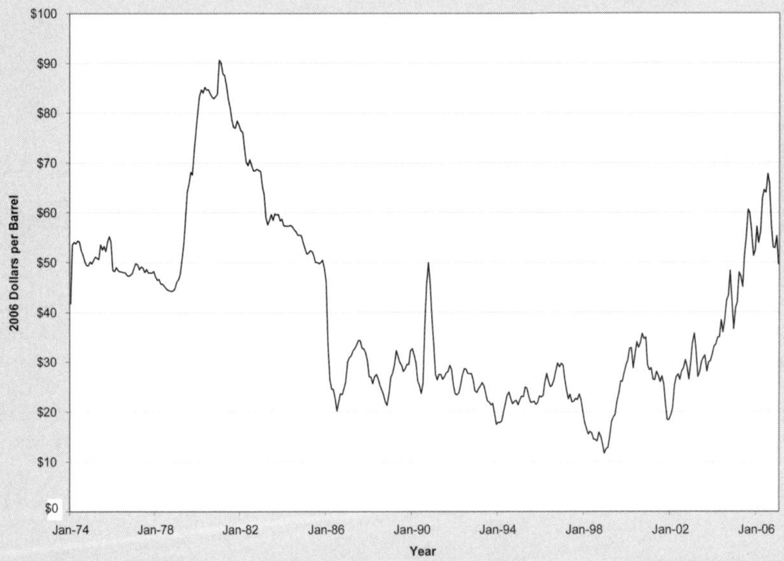

FIGURE 6-3. Inflation-adjusted crude oil price in the United States from 1974 to 2007. Prices are in dollars per barrel constant 2006 dollars.
SOURCE: Data from U.S. Energy Information Administration (2007b).

psi). All cases reflect the use of the best commercially available membrane and energy recovery technologies. Subsidies, land costs, and distribution costs have not been included, nor have the costs of concentrate management. This analysis did not use a baseline cost model of the desalination process—there are other tools available to estimate desalination costs developed by various companies and agencies in the industry, but there are limitations to each of them. Instead, the committee concluded that the transparency of actual operating data from existing desalination plants would serve as a better baseline for this sensitivity analysis. Brackish water desalination analyses were conducted where feasible and where trends were expected to differ from the seawater desalination analysis.

In this analysis, it is important to be precise about terminology. Total process costs include all of the costs incurred (or estimates of those costs) during construction and the productive lifetime of the project (exclusive of the costs of concentrate management). Capital costs include all of the costs of capital, including principal and interest. Capital costs typically do not vary during the life of the project and are treated as fixed costs. Operating costs include the variable costs of operation over time and are expressed as annual operating costs. For this analysis, it was necessary to put capital costs and operating costs on a common footing even though the former are fixed and the latter are variable. Annualized capital costs were determined by calculating the yearly capital consumption costs and interest charges as a function of different assumptions about equipment lifetime (i.e., depreciation schedule) and the cost of money (i.e., interest rate). Annualized capital costs were added to the annual operating cost to obtain the annual cost. Annual costs can be multiplied by the life of the project to obtain total project costs (exclusive of concentrate management). The annual costs per cubic meter are derived by dividing annual costs by the annual production of the plant. Most of the analysis is presented in terms of total annual costs, although a few key factors affecting capital costs are discussed separately. The sensitivity analysis was performed on these empirical data by linearizing the cost data associated with each respective variable. For example, in the analysis of the effect of membrane life on desalination cost, as membrane life was increased to 10 years from the baseline life of 5 years, costs per cubic meter contribution from membrane life were multiplied by 5/10 (i.e., 0.5). When membrane life was decreased to 1 year from the 5-year baseline, the costs per cubic meter contribution from membrane life were multiplied by 5/1 (i.e., 5). All other costs that were not affected by membrane life were held constant. All of the cost data were then normalized to the actual baseline cost data for the conventional pretreatment seawater desalination case supplied for the analysis so that the results reported are relative costs per cubic meter. The results are expressed in relative terms, so that the cost sensitivities could be more easily compared (as percent change) across the different variables and to eliminate the effects of different absolute magnitudes. All of the data from the sensitivity analysis are supplied in tabular form in Appendix C.

Impact of Water Source, Scale, and Process Design on Capital Costs

Initially, it is important to acknowledge that membrane desalination facilities exhibit very real economics of scale in both seawater and brackish water applications. Data on capital costs of seawater desalination as a function of plant size are presented in Figure 6-4. As the scale of the system is increased from 38,000 to 380,000 m^3/day (10 to 100 MGD), capital costs, expressed as cost per cubic meter of product water, decline by 20 percent. Other published data also illustrate the impact of scale on capital costs, especially as the scale is decreased below 38,000 m^3/day where the costs rise even more dramatically (USBR, 2003). When feedwaters are brackish, cost also declines as plant scale increases, albeit to a smaller degree (<10 percent).

Data also are presented to illustrate how capital costs vary with different pretreatment operations included in the total treatment system. The marginal increase in the capital costs of a system with an ultrafiltration (UF) or microfiltration (MF) pretreatment process compared to a conventional pretreatment process should be noted (Figure 6-4). The significant benefit of the UF/MF-based pretreatment is realized through reduced operating costs. As shown in Figure 6-5, the annual costs for a seawater RO system with UF/MF pretreatment are projected to be approximately 5 percent lower than one with a conventional pretreatment system. Care should be taken to account for these sorts of trade-offs between capital and operating costs.

Impact of Critical Operating Variables on Annual Costs

Annual costs, which are made up of annual operating costs and annualized capital costs, are a function of many variables. The key variables for RO include the quality of the water source (measured in terms of TDS or salinity concentration, total suspended solids, and organic foulants), the scale of the plant, the process, membrane life, operating pressures, and the cost of funds. The two largest components of annual costs in RO desalination of seawater are the cost of energy to operate the plant and the annualized capital costs, which include the annual repayment of principal and the interest payment (see Table 6-2 and Figure 6-6). The

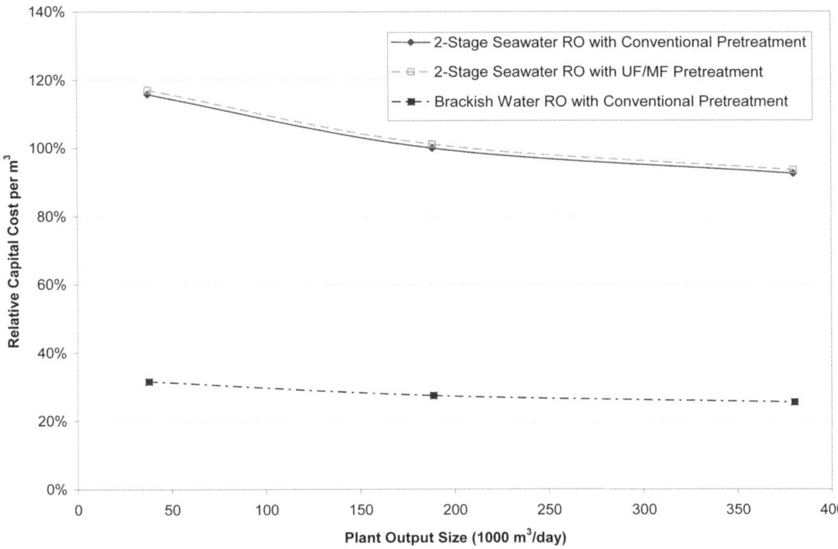

FIGURE 6-4. Relative capital costs per cubic meter for seawater and brackish water RO desalination according to facility size. Brackish water is 1,000 mg/L TDS.

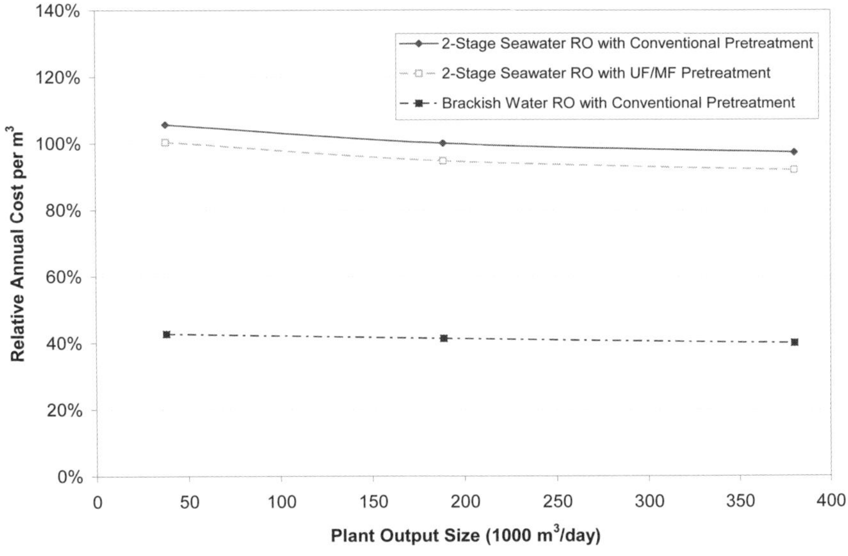

FIGURE 6-5. Effect of facility size, water source, and pretreatment process on relative annual costs per cubic meter for RO plants. The baseline assumptions for this scenario are as follows: energy costs are constant at \$0.07/kWh; membrane life is assumed to be 5 years; nominal interest rate is 5 percent; depreciation period is 25 years.

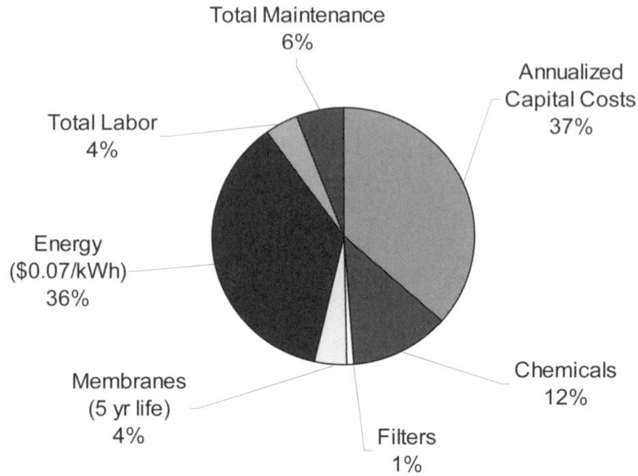

FIGURE 6-6. Annual cost breakdown in a 189,000 m^3/day seawater RO plant with conventional pretreatment. The baseline assumptions for this scenario are as follows: energy costs are constant at $0.07/kWh; membrane life is assumed to be 5 years; nominal interest rate is 5 percent; depreciation period is 25 years.

specific impacts of these variables are described in the following paragraphs.

Water Source. The annual costs of membrane desalination plants are very sensitive to the salinity content of the source water (Figure 6-5). In this analysis, energy costs, membrane life, and the cost of money are held constant. As a general rule, it will cost 50 percent more per cubic meter to produce freshwater from seawater as opposed to brackish water with a salinity of approximately 2,000 ppm. The sensitivity of operating costs to the salinity content of the source water supports the general proposition that, when concentrate management costs are ignored and all else is equal, it will virtually always be cheaper to desalinate brackish water than seawater when using membrane technologies. As discussed previously, concentrate management costs can be very high, where deep-well injection and ZLD increase the total costs of brackish water desalination by 50 to 200 percent above the desalination process costs alone (Mickley, 2007). With thermal techniques, operating costs are not sensitive to the salinity of the water. Therefore, thermal technologies are rarely used to desalinate brackish waters.

Cost of Money. It is important to recognize that, as with any capital investment, interest costs (or the costs of money) are invariably one of the larger components of total project cost. For example, the total repayment of a $1 million loan over 30 years at a nominal interest rate of 5 percent is $1.93 million. More pertinently, as shown in Figure 6-7, interest costs rise significantly as interest rates rise. An increase in the interest rate from 3 to 7 percent leads to a 15 percent increase in annual costs. Thus, the ability to secure relatively favorable rates of interest has a strong bearing on both the financial and the economic feasibility of any project. The longer the life of the project, the lower the annualized capital costs, other things equal, because the payment of principal is spread over a longer period of time. Nevertheless, the longer the project life, the higher the total costs since the magnitude of interest charges grows exponentially with time. As a matter of current practice, for this analysis the payback period is set at 25 years; however, some owner operators set more aggressive payback periods of 10 to 20 years for seawater desalination projects.

Energy Pricing. Despite recent improvements in energy efficiency (see Chapter 4), RO desalination processes are energy intensive and depend on prime energy sources. The sensitivity of annual costs (and total project costs) to the cost of energy is illustrated in Figure 6-8 for both brackish and seawater desalination. An increase in energy costs from

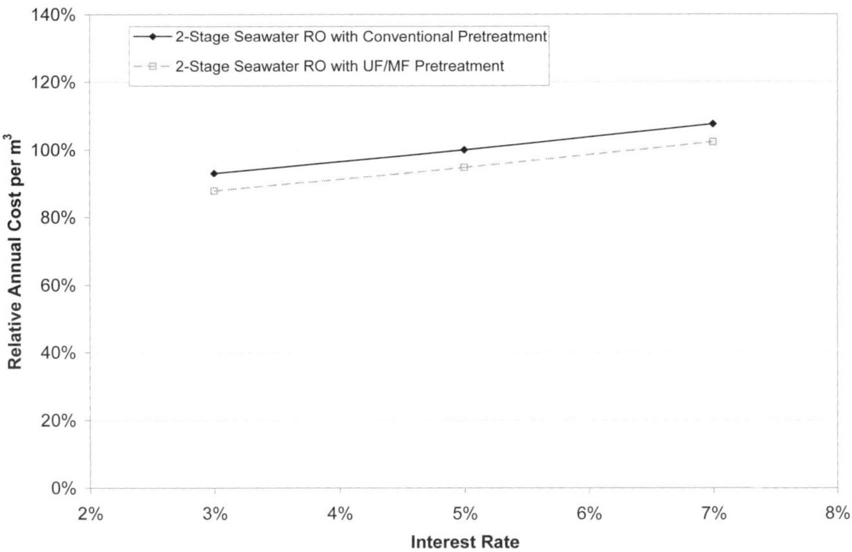

FIGURE 6-7. Effect of cost of money on relative annual costs per cubic meter for a 189,000 m^3/day seawater RO plant with conventional pretreatment or membrane pretreatment. The baseline assumptions for this scenario are as follows: energy costs are constant at $0.07/kWh; membrane life is assumed to be 5 years; nominal interest rate is 5 percent; depreciation period is 25 years.

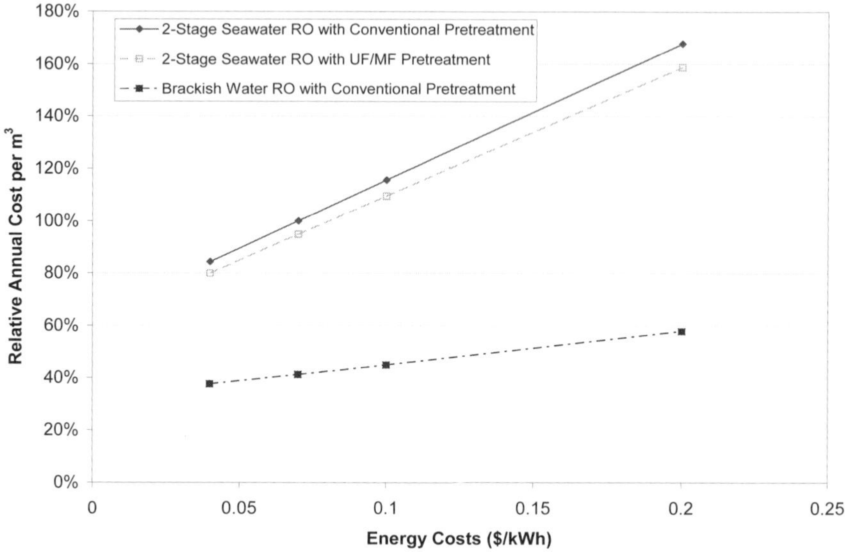

FIGURE 6-8. Effect of energy costs on relative annual costs per cubic meter in a 189,000 m^3/day seawater RO plant with conventional pretreatment or membrane pretreatment and a 189,000 m^3/day brackish water RO plant with conventional pretreatment. The baseline assumptions for this scenario are as follows: membrane life is assumed to be 5 years; nominal interest rate is 5 percent; depreciation period is 25 years. Please note the different scale for this figure.

$0.04 to $0.10/kWh results in an increase in total costs of over 35 percent for seawater desalination. Because brackish water RO desalination uses less energy than seawater desalination (see Figure 4-10), smaller increases in annual costs are observed. Efforts to reduce energy costs, as well as reductions in the total capital costs of the system, offer the greatest prospect of significant reduction in the total costs of seawater desalination systems.

Operating Pressure. Figure 6-9 shows the results of a sensitivity analysis of the effects of changes in operating pressure from a typical baseline operating pressure on relative annual costs per cubic meter, when other system variables are held constant. Although today's best available seawater RO membranes are operating at pressures that are only 40 percent greater than the osmotic pressure of seawater, it is possible that further improvements in membrane permeability and salt rejection could lead to additional reductions in operating pressures in the range 10 to 15 percent with corresponding reductions in energy usage for the seawater desalination unit operation by as much as 15 percent (see Chapter 4 and Appendix B). For a fixed energy cost of $0.07/kWh, a

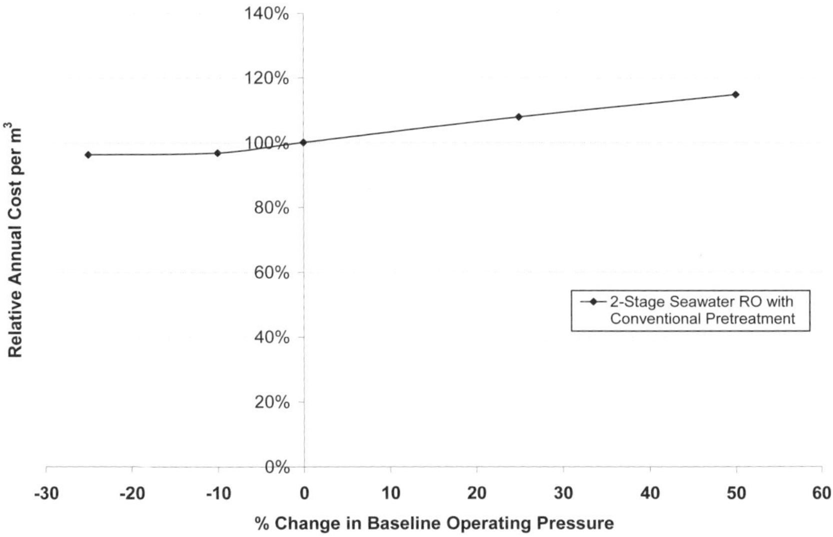

FIGURE 6-9. Effect of operating pressure on relative costs per cubic meter in a 189,000 m^3/day seawater RO plant with conventional pretreatment. Energy costs held constant at $0.07/kWh; membrane life held constant at 5 years; nominal interest rate is 5 percent; depreciation period is 25 years.

reduction in operating pressure of 15 percent could result in approximately 5 percent reduction in annual costs (Figure 6-9). At higher energy costs, such improvements in membrane technology could lead to total annual cost reductions approaching 10 percent.

In addition to permeability and rejection characteristics, the potential of membranes to foul has an important impact on treatment costs. Fouling requires increases in operating pressures if the membrane is to remain effective, and increases of 25 percent are not uncommon. An increase in operating pressure of this magnitude would increase annual costs by over 8 percent per cubic meter (Figure 6-9). At higher energy costs, the impact of fouling on annualized operating costs is even more severe and could increase costs in excess of 15 percent.

Membrane Life. The cost of membranes has fallen in recent years, and this fact is widely cited as one explanation for the increasing attractiveness of desalination. Membrane costs are now quite modest, ranging from only 3 to 5 percent of annual costs (Figure 6-6). One of the major operating issues in membrane seawater desalination plants is the issue of shortened membrane life that can result from system imbalances that lead to fouling and the need for accelerated cleaning cycles. Thus, for example, a decrease in membrane life from 5 to 3 years will increase annual

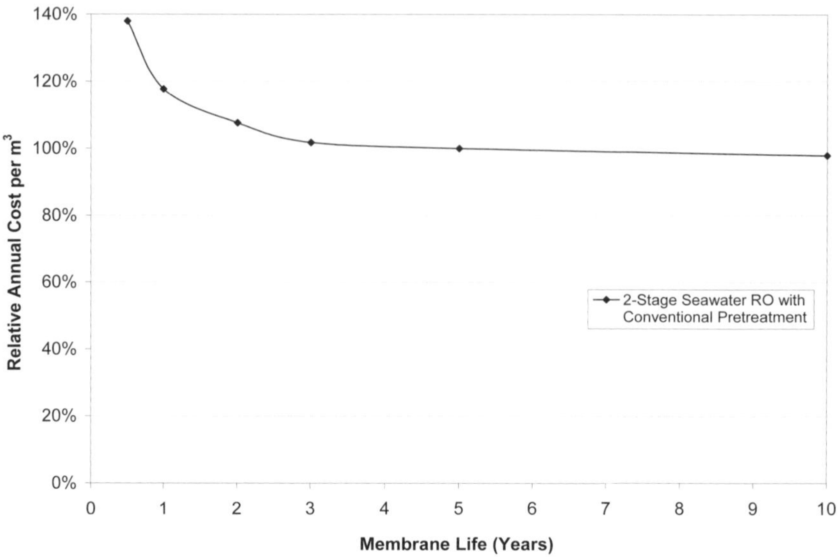

FIGURE 6-10. Effect of membrane life on relative annual costs per cubic meter in a 189,000 m^3/day seawater RO plant with conventional pretreatment. Energy cost is assumed to be constant at $0.07/kWh; the nominal interest rate is 5 percent; and the depreciation period is 25 years.

costs by over 3 percent, as shown in Figure 6-10. Catastrophic, irreversible membrane fouling leading to membrane life of less than 1 year will increase annual costs by over 25 percent.

On the other hand, for conventional systems, total project costs can be reduced by 3 percent if membrane life can be extended to 10 years over the 5-year baseline period that is common today (see Figure 6-10). Such a change would have a somewhat greater positive impact on costs for membrane-based pretreatment systems. These data and analyses, together with the fact that membrane life accounts for such a small proportion of total project costs, suggest that further research focused on extending membrane life is unlikely to have significant payoffs.

Key Determinants of Desalination Costs

An examination of the available cost data and a sensitivity analysis of the determinants of water production costs permit some general conclusions to be drawn. Five conclusions have emerged:

1. The two largest components of annual costs are annualized capital costs and energy costs. This is true of both thermal and nonthermal

technologies as illustrated in Table 6-2. Such costs are very significant, each representing over one-third of annual cost for a 189,000 m^3/day seawater RO facility (Figure 6-6). Significant reductions in the water production component of desalination costs will require significant reductions in the costs of one or both of these components. Frequently, interest charges are the largest or nearly the largest single component of cost for desalination plants. It is important to note that reductions in scale or in the capital cost of a facility will invariably reduce interest costs.

2. Development of membranes that operate effectively at lower pressures could lead to 5 to 10 percent reductions in annual costs (excluding concentrate management costs) for seawater desalination, primarily through reduced energy use.

3. Plant size is an important variable for seawater desalination projects as there are significant economies of scale in the production of freshwater. This has significant implications for small communities who are not able to take advantage of economies of scale and are therefore confronted with the higher costs of small projects. Economies of scale also exist for the production of freshwater in brackish desalination plants although they are smaller.

4. Membrane desalination costs are sensitive to the salinity of feedwater. Therefore, when concentrate management costs are ignored and other things are equal, RO desalination of brackish source water will nearly always be cheaper when compared with seawater. However, depending on the available alternatives, concentrate management costs, especially at inland locations, can increase the total costs of brackish water desalination significantly, up to 200 percent above the desalination process costs alone.

5. Lengthening membrane life from 5 years to greater than 5 years is unlikely to have a significant impact on overall costs.

Economic Costs

The costs described in the preceding section are all appropriately characterized as financial costs. However, as explained earlier, economic costs of desalination and other water supply alternatives, including traditional sources, also need to be considered. For the most part, economic costs include the external costs that are borne by the public at large that the water provider can avoid. In the case of desalination facilities, the most significant category of external costs are environmental costs. These could take several forms (see Chapter 5). The environmental costs of surface water concentrate discharges are virtually never monetized and in many instances are not even well understood from a biological perspective. Examples of other external costs that might be associated with desalination facilities include the loss of environmental amenity values because the facilities may be unsightly or interfere with lines of sight. The cost of air pollution stemming from energy generation necessary for desalination would be yet another example. Although such costs are rarely monetized, there are numerous techniques which allow them to be estimated, either directly or indirectly, with considerable accuracy. These techniques were recently summarized and analyzed by the National Research Council (NRC, 2004c).

Many external costs can be attenuated or mitigated if they are accounted for in project design and operations. Federal and state water quality and marine protection legislation normally require that the water providers consider environmental costs and take appropriate steps to minimize or mitigate them. This can require either additional capital facilities or changes in operating routines or both. The costs of these facilities and operations will appear as financial costs.

External costs are, however, rarely treated in an economically optimal fashion in the planning of desalination facilities. Most environmental quality legislation does not require a precise balancing of the costs and benefits of facilities and operations that protect environmental quality; thus, the financial costs may be equal to, less than, or greater than the true environmental costs. Often external costs are completely ignored and, where they must be considered because of laws and regulations, there is little attention to whether the external costs that are ultimately borne are optimal.

Pricing

Historically, production of desalinated water has not been a particularly attractive water supply option because the costs have been high. Ultimately, costs need to be recovered through some combination of government support and the pricing of water. While government subsidies have been a part of the picture, the pricing of desalinated water is set to defray the majority of the costs, and prices have not been competitive with those from other sources. There is evidence that this may be changing in some specific situations. Nevertheless, the pricing of water is complex and there are complicated problems that must be solved in establishing pricing structures and deciding what level of price to charge consumers. In almost every instance, trade-offs are involved. In this section, the principles and purposes of pricing are elucidated and some important practical applications are discussed.

Goods or services are said to be scarce when the wants for them exceed the supply. In these circumstances, the good or service in question is said to have value, and a method must be found for rationing the limited supply among the more or less unlimited wants. Markets are one commonly used institution whose function is to ration scarce assets, and they accomplish their tasks by generating prices and rationing the limited number of goods or services according to who is willing to pay for them. Where markets function reasonably well, the prices that emerge reflect the value of the commodity or service in question with higher prices, indicating higher value, or lower prices, indicating lower value.

Prices are very efficient forms of information and tell consumers, at a glance, whether a commodity is scarce or plentiful and what its value is. Water is not usually traded in well-functioning markets. Instead, the price is usually regulated and covers the costs of capturing, treating, and distributing the water; the water itself is frequently assigned a scarcity value or price of zero. A price of zero sends a number of important additional signals. First, it suggests that the commodity is freely available or available in limitless amounts, when in fact it is not. Inasmuch as consumers tend to make decisions about how much of a commodity to consume by consuming up to the point where the cost of the last unit consumed is equal to the benefit from the last unit consumed, pricing water at zero encourages consumers to use the water up to the point where the last unit used has a marginal value to the consumer of zero. If water is assigned a price of zero (or if water is underpriced) consumers are induced to use more water than they would if it were accurately priced, and it encourages uneconomical uses in a time of scarcity. One means of responding to water scarcity would be to charge consumers rates that include a scarcity value or a reasonable approximation thereof. Some purveyors have used price effectively as a tool of demand management. Faced with a need to reduce per capita rates of consumption, the El Paso Water Utility increased rates on a number of occasions, which led to reductions in water uses each time. The last of these involved a price increase of 35 percent, which caused a 5 percent reduction in total domestic water consumption (E. Archuleta, El Paso Water Utility, personal communication, 2006).

Historically, in most locations the cost of desalinated seawater has been very high, both in absolute terms and in relative terms compared with the cost of other alternatives. Where desalination capacity has been built in recent years, water scarcity has been high (as in Trinidad, Israel, and Perth, Australia), the value of reliability and local control has been important (as in Singapore), or certain kinds of subsidies have artificially lowered the apparent cost (such as long-term energy contracts, reduced land costs, or low-interest loans). Nevertheless, process cost reductions attributable to improved technology combined with the rising cost of most water supply alternatives suggests that the gap between the costs of desalinated seawater and the costs of alternative sources is narrowing. It is not unusual, however, for the cost of desalinated seawater to be twice the cost of existing water supplies. In spite of this fact, the addition of desalinated water is often paid by incremental increases in the price of water paid by users. The San Diego Water Authority is a case in point. Existing rates are approximately $0.24/m^3 ($300/ a.f.). The costs of seawater desalination are estimated to range somewhere between $0.64 and $1.04/m^3 ($800 and $1300/a.f.). Yet, the Water Authority estimates that if seawater desalination were used to augment the existing supply (approximately 800,000 a.f.) by a little over 10 percent (89,600 a.f.), the rate increase would be approximately $40 annually. How can the rate increase be so modest and what does it mean?

Public water providers typically price water according to the average cost of acquiring, treating, and delivering it. In some states and locales, this is required by law. Historically, this was necessary because capital-intensive industries such as water purveyors have very high fixed costs relative to operational or marginal costs. Pricing structures that recovered only marginal or operational costs would leave the operator unable to cover fully the costs of debt service and repayment of the capital costs. Today, the average cost pricing rule continues to be followed even though marginal or operating costs are now much higher than they once were. One contemporary consequence of this is that the costs of relatively high-priced increments of additional water supply are averaged in with the (much) lower costs of the existing supply. Consider, for example, a new supply that costs $0.8/m^3 ($1,000/a.f.) and an existing supply that costs $0.24/m^3 ($300/a.f.). Suppose the existing supply is augmented with an additional 10 percent of the high-cost supply. The resulting average cost—the cost of the existing and augmented supply averaged together—would be $0.29/m^3 ($363/a.f.) even though the costs of the new supply was almost three times that high. The implications of using such average cost pricing rules are mixed. On the one hand, it keeps the water rates paid by consumers from rising sharply. On the other hand, by buffering the consumer from the higher costs of the new supply, the practice sends erroneous signals about the true cost of the additional water, suggesting to consumers that water is cheaper and more plentiful at the prevailing price than is true in fact. Consumers respond by using more of the higher-cost (e.g., desalinated) water than they would if faced with the true price or costs (although total water use would be expected to decline in the face of higher prices). The result is that the high-cost supply is inefficiently used. More of that supply is consumed than would be the case if consumers were charged the true price. In contemporary circumstances, the practice of average cost pricing, then, stimulates water use artificially beyond the point where it is efficient. Stated differently, the value of water in the artificially stimulated uses is less than what it costs to make the marginal increments of supply available.

In an era when water scarcity is both pervasive and intensifying, policies and institutions that create rates and costs that are lower than the true or "marginal cost" appear perverse.[1] While it may be difficult to separate high-cost water from low-cost water in distribution systems, it is not difficult to identify the consumers who require the additional, high-cost supplies to be developed. There is no rule that requires that newcomers and long-time residents be charged the same price. The use of marginal cost pricing would raise the cost of water to users of the marginal or most recent addition to supply very sharply. However, there are concerns about the fairness of charging sharply higher prices for the new water. Most of these concerns are directed at the question of how the poor can afford to pay higher prices for water, which is, at some level, an essential commodity. In general, the benefits of marginal cost pricing can be gleaned without unduly penalizing the poor by devising rate schedules that mimic marginal cost pricing. Such schedules entail very low or zero prices for the first block or "life-line" quantity of water. This is the amount of water necessary for drinking, cooking, and basic sanitation. Thereafter, the price of ensuing increments of water rises so that the more water a consumer uses, the higher the price paid at the margin. This type of rate structure is widely used with electricity and has been adopted by many water purveyors in recent years. Fundamentally, such rate structures mimic marginal costs by pricing successive blocks of use at progressively higher rates. The intention is to discourage progressively larger quantities of consumption beyond some lifeline amount, and the evidence suggests that such rate structures are associated with lower levels of consumption when compared with the declining block rate structures (Haneman, 2006; Renwick and Green, 2000).

The true cost of water, including desalinated water, is likely to remain unstated or understated. This means that, in general, consumers will behave as if water scarcity is less intense than it is in actuality. Although it is desirable to reform rate structures so that consumers are confronted with a realistic approximation of the cost of the water they use, political and institutional inertia sometimes makes this difficult. The result will be overinvestment in desalination facilities and the continued application of desalinated water to uses whose value is less than the cost of making the

---

[1] This discussion of marginal cost pricing refers to short-run marginal costs and *not* long-run marginal costs. The difference is that in the short run (i.e., over the life span of a desalination plant) certain costs such as capital costs are considered fixed and invariant. Thus, in the short run, the marginal costs of these factors of production are counted as zero because they cannot be altered. In the long run (i.e., considering replacement plants) all costs are variable, including capital costs, so long-run marginal costs would include capital cost and other costs which are fixed in the short run.

water available. A first step toward correcting this situation would be for water purveyors to disclose fully and accurately the costs of desalinated water (as well as the costs of water from all other sources). Such disclosures should include a full accounting of all costs and report subsidies and as well as distortions introduced through regulation or market imperfections. Ideally, the local price of water would be established by the interaction of supply and demand such that the resulting price covers financial, environmental, and social costs.

Other Cost Considerations

The Role of Subsidies

The development of most desalination facilities entails public subsidies of some sort. These are usually found in research and development efforts and occasionally in construction and operation. Although subsidization has been practiced extensively in the development of water supplies, it is reasonable to ask whether such subsidies are both necessary and justified. The economic justification of subsidies relies on the notion that subsidies are justified when the benefits of the subsidized activity are widespread in nature and cannot be fully captured by the firm or utility that generates them. As a rule there will be underinvestment in such activities in the absence of the subsidy. Thus, for example, subsidies may well be justified to promote environmental protection and enhancement activities and to underwrite the costs of water quality maintenance and improvement when the benefits from such improvements are widespread. There will be many instances where research and development activities related to the development and implementation of desalination technologies can justifiably be subsidized because the developing agents cannot capture all of the returns to the scientific information that is developed when that information is freely available to others. Subsidies do not meet the standard economic tests of justification where the research and development information is held as proprietary. In some instances, subsidies are promoted in order to keep water rates down or in an affordable range. There is no scientific basis which justifies this kind of use of subsidies. Rather, they are the result of political processes and the value and equity preferences inherent in them.

Public and Private Roles

Earlier studies have concluded that, because the demographic, economic, political, and physical circumstances vary so widely, there is no single model of public or private water service delivery that is preferred in all situations (Gleick et al., 2002; NRC, 2002c). A canvas of desalination projects around the world reveals that some facilities are entirely publicly constructed and operated, some are private, and some are mixed. The presence of such a diversity of public, private, and mixed arrangements for the design, construction, and operation of desalination facilities would appear to confirm these earlier conclusions about the desirability, or lack thereof, of privatizing the provision of water and wastewater services. Indeed, it now seems to be recognized that privatization is but one of an array of approaches to acquire sufficient capital as well as operational experience with water and wastewater services of all types (Raftelis, 1989).

It is important to recognize that there is a range of privatization options. They include (1) private provision of services (e.g., laboratory work) and supplies (e.g., chemicals); (2) private contracting for the operation and maintenance of a capital facility; (3) negotiated contracts for the design, construction, and operation of new facilities; or (4) sale of water utility assets to a private firm. The first three options are found throughout the global desalination industry and while there are currently no examples of the fourth without the government retaining some ownership, as in Aruba and Malta, for example, it does remain an option. Each of these options is known to work well in some circumstances. It is, however, difficult to generalize about such circumstances (NRC, 2002c). A more detailed discussion of various common public-private partnership models is provided in Chapter 7.

One circumstance in which privatization may prove attractive is the provision of water supply and wastewater treatment services for smaller communities, which tend not to have the resources to take advantage of economies of scale and which may lack the resources to acquire the necessary scientific, technical, and financial expertise. Opportunities for "regionalizing" utility services across a number of communities may not always favor private provision of those services, however. In general, in selecting arrangements along the public-private continuum, it will be important to recognize that private and public agencies face different kinds of incentives and have different strengths. For example, public agencies tend to be more responsive to political currents and tend to do a better job of providing public goods such as environmental protection (Baumol and Oates, 1988; Oakland, 1987). On the other hand, private entities are thought to operate more efficiently, although they are frequently subject to regulation (NRC, 2002c). There is reason to believe that these conclusions about public and private provision of water and wastewater services will apply to desalination. Indeed the existing diversity of public and private arrangements found around the world lends strong support to the proposition.

THE STRUCTURE OF BENEFITS

The benefits of desalination accrue in terms of the value of the water produced. Where markets function well, market clearing prices tend to be good indicators of value and can be used to compute the value of a fresh, reliable water supply to consumers in the various water-use sectors. The difficulty in computing the benefits of water supply in this conventional fashion lies with the fact that, as discussed previously in this chapter (see Pricing), water is rarely priced in markets. As a practical matter, the value of water produced from desalination facilities is frequently established with reference to the costs of acquiring water from the least costly alternative source. At its simplest, if the cost of desalinated water is $0.8/m^3$ (\$1000/a.f.) and the next most attractive option costs $1.2/m^3$ (\$1500/a.f.), then the value of the desalinated water would be counted as $0.4/m^3$ (\$500/a.f.). This illustrates why the value of the water cannot be established in the absence of reference to the costs of acquiring water from other sources. Unfortunately, the problem of valuation is often not usually so simple. The qualities of water from different sources differ; the reliability of water from different sources differs; and the environmental costs associated with different sources differ. In some instances desalinated water can be substituted for other water supply sources that are not being exploited in a sustainable fashion (e.g., persistently overdrafted groundwater; see Box 3-3), and benefits accrue to the provision of desalinated water in the form of reduction in unsustainable uses of alternative groundwater and surface water sources. An accurate valuation of desalinated water would include a premium on quality because such waters are likely to be of the highest possible quality and have a premium on reliability since desalination processes (particularly those that use seawater as a source) are uncoupled from the hydrologic cycle and, thus, are among the most reliable of water sources.

Value of Reliability

Water supply reliability can be defined as the consistent availability of water in response to demand. Reliability of supply, for example, is most obviously manifested by the fact that water flows from taps in most U.S. homes when they are turned on. A reliable water supply is more valuable than one that is susceptible to interruption. Indeed, water purveyors are willing to pay a premium for water that is a reasonably accurate indicator of the value of reliability. This value and the premium depend on a number of factors, including the use to which the water is put, the availability of alternative sources, production costs, and the costs of a supply disruption. The importance and implications of these factors will differ from region to region and, thus, the value of reliability will differ from case to case (Cooley et al., 2006).

One of the most desirable attributes of desalination is the fact that the availability of brackish estuarine waters and seawater is independent of the hydrologic cycle. This means that the capacity of coastal desalination projects to produce water is not affected by severe droughts, an attribute that is particularly valuable in circumstances where climate is highly variable. Brackish groundwater desalination can also provide reliable water supplies during short-term droughts, although longer periods of drought can affect regional groundwater availability. This reliability will also be significant in arid and semi-arid areas where rainfall and runoff are often inadequate to serve existing demands, where water is overallocated, and where water allocations, water rights, and the ability to access and use other sources of water are in dispute.

Purveyors are willing to make substantial investments to avoid supply interruptions, often because the cost of such interruptions may be larger than the cost of improving the water supply reliability. Thus, for example, the East Bay Municipal Utility District (EBMUD) invested more than $200 million to protect reservoirs, pipes, pumps, and treatment plants against a large earthquake. EBMUD serves over one million consumers on the eastern margins of San Francisco Bay, and it is estimated that in the absence of such strengthening, water-related losses to consumers from a large earthquake on the EBMUD system could be as high as $2 billion (EBMUD, 2005). Earthquakes are not the only potential threats to reliability. Other threats to reliability include water quality contamination, climate change and its impacts on quality and water availability, changes in the regulatory environment, and price volatility in the various factors needed to operate the system, such as energy. Unanticipated threats such as the renewed potential for terrorist attacks can also arise at any time.

Water supply reliability is measured in different ways. The most common measure is to portray the risk of a projected supply falling below a projected demand over some specified time period. When a system is described as having a reliability of 95 percent, that implies that supply will be equal to or exceed demand in 19 instances (e.g., days, months, or years) out of 20. Other approaches depict the severity of water shortfalls, and this should always be taken into account. A system with a reliability of 90 percent might be more attractive than one with a reliability of 95 percent if the shortfalls in the former supply are very small and those associated with the latter are less frequent but much larger (Cooley et al., 2006).

There are a number of alternatives available for improving supply reliability. For example, dams and reservoirs and the associated conveyance facilities tend to insulate consumers from short-term variations in precipitation and runoff by storing water during wet periods for use in dry times. Multiple sources of supply also augment system reliability, particularly when the different sources are hydrologically independent of each other or provide a means of buffering against nominal hydrologic variability. No source is completely invulnerable to supply disruption. For all its reliability attractiveness, desalination technology is especially vulnerable to adverse changes in the energy supply picture, for example.

Where reliability is valued it should be counted as a benefit of desalination technologies. There is frequently an issue of whether, in fact, a more reliable supply justifies higher values and how reliability ought to be valued. If water were traded in well-functioning markets, it would be relatively easy to answer this question by examining the difference in prices as between a more reliable supply and a less reliable supply. The absence of market-generated prices means that techniques for imputing the price must be employed in order to value reliability. One means that is frequently used by economists is to evaluate the willingness to pay (WTP) of consumers to reduce the probability of a water shortage. Suppose, for example, that the risk of a water shortfall that would require rationing is 1 chance in 50. What would consumers be willing to pay to reduce that chance to 1 in 75? Economic studies have shown that the WTP to avoid restrictions on water use due to drought or other factors ranges from \$32 to \$421 per household per year in constant 2003 dollars (Carson and Mitchell, 1987; Griffin and Mjelde, 2000). When the estimated reduction in water use due to drought is multiplied by the probability of the drought actually occurring, these estimates imply that the reliability value of extra water in severe drought circumstances could be as high as \$3.12/m^3 (\$3900/a.f.) (Raucher et al., 2005). Although there may be controversy surrounding the magnitude of this figure, the study results do indicate that greater water supply reliability has value.

There are other techniques for estimating the value of reliability. One, developed by the Pacific Institute, is described in Appendix D. The method borrows from and adapts tools from financial portfolio theory and allows a comparison of water supply alternatives that have differing degrees of reliability.

CONCLUSIONS AND RECOMMENDATIONS

Historically, the relatively high financial costs of desalination constrained the use of desalination technologies in all but a few very specific circumstances, but the cost picture has changed in a number of important ways. There have been significant reductions in membrane costs and in other components of cost in the production of desalinated water. Perhaps, more significantly, the costs of other alternatives for augmenting water supplies have continued to rise along with the degree of treatment required of existing supplies, making desalination costs more attractive in a relative sense. A continuation of these trends would likely make desalination costs more attractive and less of a constraint in the future. The trend of desalination process cost reduction may be abetted through a program of strategically directed research aimed at achieving potentially large cost reductions. The following recommendations are based on the detailed discussion and analyses of this chapter.

Substantial reductions in the financial cost of desalination will require substantial reductions in either energy costs or capital costs. Energy and capital costs are the two largest components of financial cost for both thermal and membrane seawater desalination processes. It is important to recognize that reductions in scale or in the capital costs of a facility will have associated reductions in interest costs. In most instances, interest costs will be a large component of total costs. Future trends in energy costs will also be important inasmuch as significant increases in energy prices could offset or more than offset cost reductions in other areas and make desalination technologies less attractive.

For brackish water desalination, the costs of concentrate management can vary enormously from project to project and may rival energy and interest costs as the largest single component of cost. The high cost of concentrate management at some inland locations ultimately offsets the cost advantage that can be obtained from utilizing feedwaters with lower salinity.

There are small but significant efficiencies that can be made in current membrane technologies that will reduce the energy needed to desalinate water and therefore offer potentially important process cost reductions. Today's best available seawater RO membranes are operating at pressures that are only 40 percent greater than the osmotic pressure of seawater and therefore are approaching the theoretical limits of energy efficiency for membrane desalination. However, development of membranes that operate effectively at lower pressures could lead to 5 to 10 percent reductions in annual costs of desalinating seawater associated with a 15 percent decrease in energy use.

Extending membrane life is likely to have a very small impact on desalination costs. Today's best-available seawater RO membranes routinely operate for 5 or more years before needing to be replaced. The ability to extend membrane life past 5 years to 10 years will have a minimal impact on total costs given the small contribution of membrane replacement costs to total costs over a 5-year lifetime. However, the prevention of catastrophic failure is especially important because membrane failure within the first year of operation can cause an annual cost increase of over 25 percent. Future research efforts should be focused on mistake-proof, robust prefiltration to ensure against premature failure of the RO membranes.

The costs of producing desalinated water have fallen in recent years but may rise in the future if the price or cost of energy rises faster than cost decreases from technological improvements. Increases in energy costs lead disproportionately to increases in desalination costs and in the costs of transporting water long distances. The ultimate size of these increases, however, may be limited when the costs of fossil fuels reach the costs of other energy technologies, especially renewable energy technologies that can substitute for fossil fuels. Consequently, energy costs will not rise indefinitely even if possible fuel prices do rise more or less indefinitely. In considering the implications of increasing energy costs, it is important to recognize that alternative supply measures that also have high energy demands will be sensitive to future energy prices.

Conservation and transfers from low- to high-valued uses will usually be less costly than supply augmentation schemes, including desalination. In many circumstances, low-cost methods of demand management could provide significant water savings. Low-cost demand-management techniques have not been exhausted and, so long as potential remains, demand management will offer the possibility of freeing up water to serve new uses at lower cost than desalination. Similarly, market-like transfers of water can also offer relatively low-cost ways of acquiring additional supplies of water. This is particularly true where additional water supplies are needed to support urban growth and where agricultural water is available for reallocation. Conservation and efficiency improvements that reduce the total demand for water often come with associated benefits (such as reduced energy costs), require little capital investment, and can be implemented relatively quickly. Ultimate costs will vary depending on the local details of water use, water available for transfer, previous efforts to improve efficiency, financial perspectives, and institutional factors that encourage or discourage different water policy choices.

To make the true costs transparent, the economic costs of desalination should be accounted for and reported accurately. Failure to price water accurately can lead to inefficient use and overuse. Melded pricing or average cost pricing is frequently used pursuant to law or to address equity consideration. This practice understates the cost of desalinated water to the consumer, and the supplier should take care in publicly reporting the true and accurate economic costs.

7

Implementation Issues

Water providers throughout the world are concerned about meeting potable water supply needs. Increasing water demands as a result of greater environmental awareness and localized population growth has placed a strain on traditional freshwater sources in some areas. Consequently, many providers are considering alternative water supply sources, including the desalination of seawater, brackish water, or both. In addition, they are pursuing advanced treatment technologies to deliver drinking water to meet more stringent drinking water quality standards.

This chapter provides a summary of implementation issues, including institutional matters, that need to be addressed by water providers when developing desalination projects. The issues covered in this chapter include environmental regulatory requirements, capital and operating costs, public perception, siting considerations, planning and design issues that consider the uncertainty of new technology, product water quality changes within conveyance systems, alternative project delivery and procurement methods, and project financing. Environmental issues, including environmentally protective design, predictive modeling, monitoring, and energy issues, are also important utility considerations, but these are not included here because they are addressed in Chapters 4 and 5.

ENVIRONMENTAL REGULATORY ISSUES

Environmental issues relating to source water and concentrate management are significant considerations for desalination projects. Environmental issues are described in detail in Chapter 5, and the regulatory aspects of desalination project implementation are discussed in this section.

The types and complexity of permits required for a desalination plant vary depending on the project location and other site-specific factors, such as the type of desalination technology and the method of concentrate management employed. The implementation of a desalination project typically requires multiple permits from federal, state, and local agencies. In general, the regulatory programs and associated permitting processes revolve, and can be broadly classified, around the three streams involved in the process (Stratus Consulting, 2006):

1. **Source water** (or feedwater stream) permits address the location and means of obtaining the source water used by the desalination facility.
2. **Potable water** (or finished water stream) permits address the use of the finished water produced by the desalination facility.
3. **Waste** (concentrate and other associated waste stream) permits address the treatment or discharge of the waste streams, including concentrate, chemical wastes from cleaning processes, and any other waste associated with the operation of the facility.

Other required permits (e.g., building, site work, roadway crossings) are similar to those required for construction of other types of water treatment facilities and are not addressed here. Some state and local authorities may require other permits in addition to those discussed in this section (see Cooley et al. [2006] for examples of California permit requirements).

Of the three categories just defined, the regulatory issues related to the permitting of the concentrate and other waste streams are typically the most involved. The key federal permit requirements are related to the Clean Water Act. To obtain the permits, extensive environmental impact analyses may be required, depending on the specific discharge method proposed. Most of the permits provide for extensive review and comment from resource agencies and the public.

Source Water Permits

Source water permit requirements depend on the location of the desalination facility. For an inland groundwater facility, no significant regulatory approval is required for the groundwater wells, unless water rights or pumping permits are required. For a stand-alone coastal desalination facility, the following permits will be required for a new intake pipe:

- a Clean Water Act (Sections 316b and 404) permit, issued by the U.S. Army Corps of Engineers, which regulates intakes and the discharge of dredged materials into navigable waters (see also Box 5-1);

- a Rivers and Harbors Act (Sections 9 and 10) permit, issued by the U.S. Army Corps of Engineers, which regulates the construction of any structure or work within navigable waters; and
- in some states, a permit from the state coastal agency.

For a coastal desalination facility co-located with a power plant, a permit for constructing a new intake (or discharge) pipe is not required because the existing intake (and discharge) of the power plant is used. Although this is an advantage of co-location of desalination and power plants, there are also disadvantages of co-location, which are addressed later. In some cases, plants have also been required to obtain separate permits to use existing intake or discharge structures.

Potable Water Permit

The potable water permit, as required by the Safe Drinking Water Act, is typical of permits required for any drinking water treatment plant. This permit is not required if the desalted water is used for nonpotable (e.g., irrigation) purposes. The potable water permit requires periodic compliance monitoring. One unique aspect of this permit for desalination is the need to identify the monitoring points in the treatment process for filtration efficiency and turbidity compliance. In most states, state agencies have primacy to regulate the Safe Drinking Water Act.

Concentrate Management Permits

The effective management of the desalination concentrate stream is a significant issue in the implementation of a desalination facility. Concentrate management (especially for inland or brackish water applications) is often the decisive factor that determines the viability of a desalination project because of both environmental and economic concerns. Specific concentrate management approaches and associated environmental issues are discussed in Chapters 4 and 5. Although several concentrate management methods are currently available, as shown in Table 4-4, there are permitting challenges, high costs, and other limitations associated with all methods (Sethi et al., 2006a). Here, the key implementation challenges and regulatory issues related to concentrate management are discussed.

Currently there are multiple levels associated with the regulation of concentrate management, including federal, state, and often local agencies with specific requirements. At present, concentrate from the desalination process is regulated through a default classification as an industrial waste under the Clean Water Act because the Act does not specifically address by-products from drinking water treatment plants. However, in the State of Florida, concentrate has been given some regulatory distinction, as it is now called a "potable water byproduct" if produced by plants of size 189 m^3/day (50,000 gal/day) or smaller. Pending state legislation may extend this to plants of larger size (Mickley, 2006). Nationally, separate classification of drinking water treatment plant by-products would require an amendment of the Clean Water Act.

The federal laws associated with the management of concentrate and associated wastes from desalting plants include the following:

- Clean Water Act. A National Pollution Discharge Elimination System (NPDES) permit is required for surface water discharges. Discharge to sewers (i.e., indirect discharge to a municipal wastewater treatment plant) does not require an NPDES permit, but compliance with Environmental Protection Agency Pretreatment Control Program standards and state pretreatment programs may be required. Engineering studies may be required by states for reuse-based concentrate management methods (e.g., land application, spray irrigation).
- Safe Drinking Water Act. Compliance with the Underground Injection Control program and with Wellhead Protection Program regulations is required for deep-well injection.
- Resource Conservation and Recovery Act (RCRA). Although the by-products of desalination plants are typically not considered hazardous, it is the utility's responsibility to confirm if the concentrate produced meets the RCRA definition of a hazardous waste.
- Solid Waste Disposal Act. This law applies to nonhazardous solid waste disposal and would apply to desalination plants using a solid waste disposal method.
- Comprehensive Environmental Response, Compensation, and Liability Act. This law is applicable only if the desalination plant has stored, treated, or disposed of a hazardous waste as defined by RCRA. This law might apply to desalination concentrate from groundwater that contains high levels of toxic elements exceeding drinking water standards.
- Hazardous Materials Transportation Act. This law applies if any hazardous residuals (e.g., cleaning waste) are transported offsite.
- Toxic Substances Control Act (TSCA). This law, which controls the sale of toxic chemical substances, applies if concentrate is defined by the TSCA chemical inventory as toxic and sold for reuse (e.g., blended with treated wastewater for reuse).

- If the waste contains technologically enhanced naturally occurring radioactive materials (TENORMs) exceeding certain levels, disposal or storage may require additional permits. Numerous state and federal regulations govern the disposal of waste that contains radionuclides, although there are currently no federal regulations that specifically address TENORMs (EPA, 2005).

The compliance process for all of the aforementioned permits is complex and necessitates a detailed review in the planning phase. Overall, the costs and time involved in the permitting process are significant and, regardless of the capacity of the facility, the regulatory requirements involved in the process are approximately similar. Depending on the concentrate management alternatives available, small inland facilities could be especially challenged by the implementation and permitting of concentrate management processes.

PUBLIC PERCEPTION

Successful implementation of a desalination plant requires more than a successful resolution of technical issues. Affected persons in the area are often able to slow or block implementation if public perception is negative, whether or not a concern is justified in the particular project. Public concerns about desalination vary and include worries about the cleanliness of the source and product water, technical feasibility of desalination, environmental effects of process operation and concentrate management, privatization issues, and future affordability of the resource. The perceptions and concerns about desalination may be influenced by the need or urgency for an additional source of freshwater or for a reliable source of water.

Failure to gain public acceptance can derail the most essential and feasible desalination project. Local citizens and nongovernmental organizations may influence a regulatory body or local government officials, and these regulators or officials can in turn place impediments in the permitting process. Broad-based public participation in the process—that is, greater than that necessitated by permitting requirements—may help minimize adverse relationships and help the project progress more readily toward successful implementation (Burroughs, 1999; Roberts, 2004; Robinson, 2007).

The following section describes a number of concerns that have been voiced by citizens in response to proposed desalination facilities. Some of these issues may not be valid on a technical level; for others, there may be ways to mitigate the concerns. Nevertheless, the public perception of a concern will need to be addressed to ensure successful implementation of the desalination project.

Source Water Issues

Desalination utilizes source water not previously considered suitable as a source for drinking water. For example, seawater contains much higher concentrations of many chemical species, such as boron, than are found in conventional drinking water supply sources or finished waters. Estuarine waters are often the receiving water for known point and non-point discharges. Inland brackish waters can be perceived as being less pristine than other groundwater. Therefore, concerns may arise among the public over the ability of the desalination process to fully treat these source waters. Because reverse osmosis (RO) and nanofiltration technologies may allow passage to the product water of some constituents in the source water (see Chapter 5), the public may perceive water produced by membrane-based desalination as not sufficiently protective of public health. Water providers can address this concern by educating the public on the technical advances of water treatment processes and the effective constituent removal efficiencies of these processes.

Environmental Issues

As discussed in Chapter 5, there are varied environmental concerns related to desalination, and these concerns can affect public perception of a desalination project. For seawater desalination, as with any process involving the intake of large volumes of surface water, there can be concerns with impingement of larger organisms and entrainment of smaller ones in the intake system. Citizens may also be concerned about adverse effects of desalination concentrate discharge. The energy requirements of seawater desalination can lead to public concerns over increased greenhouse gas emissions that can be a part of increasing energy usage (unless non-greenhouse-gas-emitting energy sources are used to support the plant or unless carbon credits are established). Public concern over global climate change can have a localized influence in limiting new sources of emissions. Water providers can address concentrate discharge concerns by educating the public about the advanced environmental models used to predict environmental changes due to concentrate discharge as well as the environmental monitoring plans that will be in place to detect any unacceptable environmental changes in the very early stages should they occur.

Necessity of Supply

Proposals for a new desalination plant can lead the public to question the need for additional water supply. If the plant supplements current water supplies, the perception can be created that the provision of additional water supply may lead to additional population growth, creating the impression that desalination would work against existing or potential growth management efforts. If current water supplies are sufficient to meet human demands for water, the desalination facility could be seen as unnecessary even if it could replace another unsustainable water source in current use or if it would allow for currently unmet environmental water needs to be fulfilled. Water providers can address the replacement scenario by educating the public on the need to replace unsustainable sources or sources that are causing unacceptable environmental impact. An explanation of the anticipated net environmental benefit can be helpful to address the public concern regarding the perceived environmental impacts of desalination projects.

Reliability of Supply

Consumers expect consistent availability of a sufficient quantity of high-quality water on demand. Although seawater desalination offers the promise of a drought-resistant water supply, the limited experience in the United States with large desalination plants can affect stakeholders' confidence in the reliability of desalination technology even though similarly sized plants are common in other countries (Robinson, 2007). Seawater desalination can encompass more unit subprocesses (e.g., filtration, chemical treatment) than can traditional water treatment, opening the probability for more processes to go wrong unless appropriate parallel steps, redundancy, and process isolation techniques are in place to mitigate potential problems. With mitigation steps in place, desalination plants can be as or more reliable than traditional water treatment plants. Nevertheless, the public may be more comfortable with the reliability of familiar treatment processes using traditional sources.

Water providers can address this concern by educating the public on the number of successful international seawater desalination projects and U.S. brackish water desalination projects that are in operation. In the United States, brackish water membrane desalination is becoming a common water treatment technology to meet water supply needs.

Energy and Cost Concerns

The water sector is a large consumer of electrical power, and desalination processes require more energy to operate than traditional water treatment processes (Table 5-1) . Thus, public concern over the reliability and availability of energy can spill over into concern for these higher-energy processes. In some parts of the country, the current energy grid capacity is already strained, and the perception can be created within the public that the system may not sustain increased demand from desalination facilities, thus increasing the potential for power outages or other failures affecting more than just the water supply.

Energy costs are a portion of the question in consumers' minds regarding the long-term cost of water produced by desalination. If energy prices rise, the operating cost for desalination will necessarily increase unless long-term power purchase agreements are in place. Thus, the public may have concerns about the future costs of the water supply. Water providers can address this concern by educating the public on the incremental cost of desalination as compared to other water supply alternatives. The public can also be informed that total project cost (including operating cost) will also be considered as part of the criteria for project selection.

Siting Concerns

Public perception can be focused on highly localized issues associated with the siting of a desalination plant. Localized environmental degradation, co-location with power plants, barriers to beach access, and increased population growth and regional development are examples of concerns voiced by citizens about desalination plants. Although some of these issues would arise with any development in these areas, there are certainly unique siting concerns for a new desalination plant. It was found in Tampa Bay, Florida, that consumer interest, positive or negative, was not strong regarding the desalination project until specific potential sites were chosen. In Tampa, some of the negative public reactions derived from the plan to co-locate the plant with an existing coal-burning power plant, reflecting citizens' displeasure with the possibility of prolonging the operational life of the power plant. The public also expressed concerns about environmental impacts on Tampa Bay (Robinson, 2007). Water providers can address these concerns by involving the public early in the siting process, clearly identifying site-selection criteria with public input, and inviting alternative site suggestions from the public.

Public and Private Water Management

Public versus private management of water supply systems has spurred public interest for a number of reasons. The key distinction between the two options is that private entities have a profit interest in the operation of the facility whereas public ownership and operation does not. Hence, the question can be raised whether a private entity will properly prioritize the quality of operation over profit. In balance, there also is a perception that private entities are more cost conscience and may be able to produce comparable water quantity and quality at a lower cost than a public water provider. Consequently, there is no hard-and-fast conclusion that the public prefers either approach (NRC, 2002c).

SITING CONSIDERATIONS OF CO-LOCATION

When planning for desalination, decisions must be made about where to site the plant, and water suppliers should consider the advantages and disadvantages of co-location. The co-location concept, which involves direct connection of the desalination plant intake and/or discharge facilities of an adjacently located coastal power plant, can bring economic and environmental benefits (Voutchkov, 2004, 2005). In the case of an inland location, the desalination plant can be co-located with a wastewater treatment plant. The following section focuses on issues that water suppliers should consider with respect to co-location.

In the case of an inland co-location of a desalination plant with a wastewater treatment plant, the concentrate can be discharged directly to the wastewater plant or blended with the treated wastewater effluent prior to surface water discharge. The former strategy requires that the impacts on the wastewater treatment processes, if any, be within acceptable limits. Another possible benefit includes using the waste gas produced during wastewater treatment as a source of energy for the desalination process.

Co-location with a power station was used for the Tampa Bay Seawater Desalination Plant and has been considered for numerous plants in the United States and worldwide, such as the proposed seawater desalination plant in Carlsbad, California. At the Tampa Bay Seawater Desalination Plant, the intake and discharge are connected directly to the cooling water discharge outfalls of the Tampa Electric Big Bend Power Station (Figure 7-1). The cooling water discharged from the condensers is 3 to 8°C (5 to 15°F) warmer than the ambient source ocean water. This is a significant benefit because the RO process requires approximately 5 to 8 percent lower feed pressure when the influent seawater is an average of 6°C (10°F) warmer. Therefore, co-located plants use proportionally

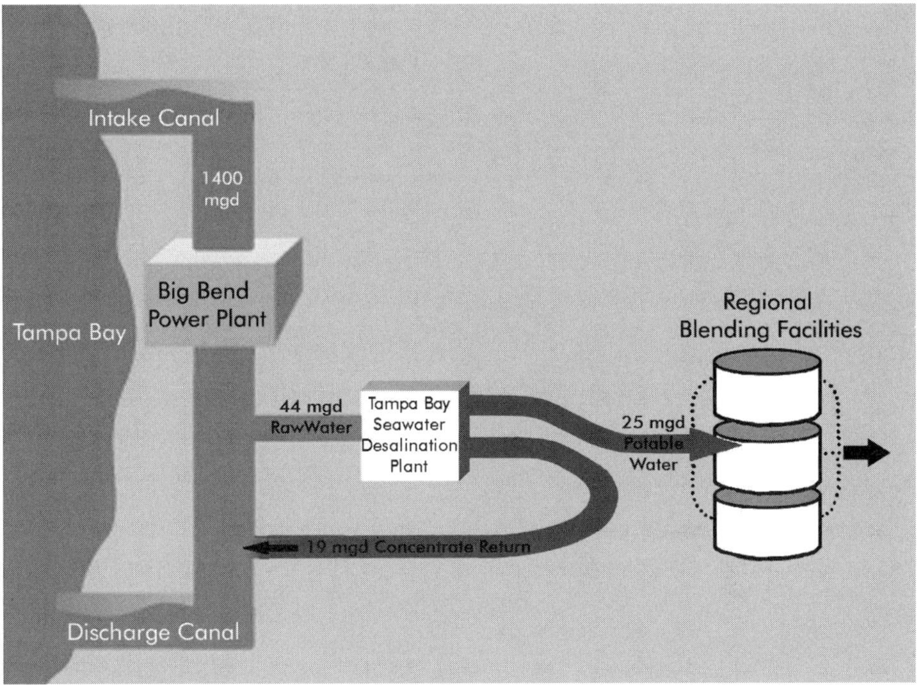

FIGURE 7-1. Schematic of a co-located seawater desalination facility.

lower energy costs for seawater desalination. However, increased temperatures in the feedwater can result in an adverse increase in salt passage and potentially accelerated biofouling of the membrane.

The source water for a desalination plant co-located with a power plant is the cooling water discharge, which has already passed through screens similar to those used on surface water intakes for desalination plants. Therefore, a co-located desalination plant generally does not require the construction of a separate intake structure, intake pipeline, or screening facilities (i.e., bar-racks and coarse screens), and co-location can also alleviate the need to construct separate ocean outfalls. As indicated earlier, a permit for a new intake pipe (or discharge pipe) is generally not required for a coastal desalination facility co-located with a power plant. The cost of intakes and outfalls for a desalination plant is about 7 percent of the total capital costs (GWI, 2006a); thus, power plant co-location yields significant permitting and construction cost savings. As a result of co-location, the grid transmission portion of the power fees can also be minimized or avoided, although state and federal regulations may sometimes prohibit preferential pricing for co-located facilities (Cooley et al., 2006).

For co-location to be cost-effective and feasible, it is necessary that the power plant discharge flow be larger than the desalination plant capacity. In cases when the concentrate is blended with cooling water, both the power plant capacity and the required "concentrate blending factor" (assessed for each particular co-location scenario) determine the maximum size of the desalination plant. Consistent with intake and outfall considerations at stand-alone plants, the power plant outfall design needs to avoid recirculation of concentrate to the desalination plant intake.

As discussed in Chapter 5, co-location can yield environmental benefits for seawater plants. Co-location can reduce impacts on marine benthic and seashore habitats relative to a stand-alone desalination plant by avoiding construction of new intakes. As long as the power plant is operating and using seawater to cool the plant boilers, the desalination facility does not cause increased impacts from impingement and entrainment. By mixing the concentrate discharge with power plant cooling water, the combined outfall will dilute the concentrate and can accelerate the dissipation of thermal and saline discharges to the ocean (Voutchkov, 2004). These benefits, however, presume the continued operation of power plants with once-through cooling systems. Therefore, the perspective and requirements of regulatory agencies are important factors for water suppliers exploring the co-location of a desalination facility. For example, the California Coastal Commission has considered phasing out power plants with once-through cooling due to the impingement and entrainment associated with screening and processing large quantities of seawater. In the absence of co-location with power plants using once-through cooling, seawater desalination facilities will have to develop other approaches to concentrate discharge, such as implementation of offshore diffuser technology. Should the power plant discontinue operating on an interim or permanent basis, water withdrawals would have to continue to provide source water and, in some cases, dilution of the concentrate. Thus, there is concern that co-location of desalination facilities might encourage the extended use of older power plants that are based on once-through cooling. Due in part to such factors and concerns, the California Coastal Commission requires that any proposed co-located desalination plant should present an analysis for the co-located scenario and for a co-located facility operating independently of the power plant (i.e., during temporary or permanent cooling system shutdowns) (Seawater Desalination and the California Coastal Act, 2004). In environmentally sensitive areas, alternate intake structures that minimize the effects of impingement and entrainment may need to be considered (see Chapters 4 and 5), especially for proposed co-located desalination plants where there are plans for phasing out the use of once-through cooling systems.

A co-location configuration introduces additional implementation complexities that should be considered along with the stated advantages. Co-location with a power plant also requires close communication between the power plant and desalination plant operating staffs. Changes in the power plant operation can affect desalination plant influent water temperature and concentrate blending ratios. Depending on the amount of cooling water being passed through the power plant, desalination production may have to be adjusted to maintain NPDES permit compliance for concentrate dilution and discharge.

PLANNING AND DESIGN WITH NEW TECHNOLOGIES

Public health protection and reliability are high priorities for public water supply facilities; therefore, utilities tend to be conservative, creating challenges in the implementation of new technologies that are unfamiliar or unproven. Although desalination has been developed successfully in other parts of the world, large-scale development of seawater desalination has not occurred in the United States with the exception of the Tampa Bay Seawater Desalination Plant. Technical problems with the Tampa Bay project have created some reluctance in other utilities to pursue seawater desalination technology. Until the Tampa Bay project or another large-scale desalination project is considered complete and operating on a sustained basis, this reluctance will likely remain.

Coastal water providers can, however, take advantage of the opportunities to learn from successful large-scale seawater desalination projects that have been developed outside of the United States. Project-specific variations (e.g., salinity, temperature, scope of work, local regulations and business practices) ensure that no two projects are identical. However, on a technical basis there are many international projects operating with more challenging feedwaters and ambient conditions and in much larger capacities than even the largest under consideration for the United States. International references and experience should be recognized as a potential resource from which domestic projects may draw.

Even with a large body of international experience, pilot or demonstration projects remain essential in desalination project planning to assess the interactions of various processes within the treatment train. Pilot testing is an iterative process, during which tests are performed and data collected to optimize the system design and operations for project-specific conditions. Pilot testing is also common for surface water and wastewater treatment systems for exactly the same reasons. Pilot testing may be used to optimize only the pretreatment steps or it may model the full treatment train including the desalination process. In membrane desalination projects, the selection and integration of pretreatment processes are key to efficient and effective operation of the treatment system. In certain cases for seawater plants, pilot studies are used to test new membranes with specific characteristics, such as greater boron or bromide removal, to meet local requirements. These pilot studies should lead to more accurate cost estimates and should greatly reduce the risks associated with the project. Pilot testing is a necessary part of the planning and implementation process and is often considered a requirement during the regulatory approval process in states where the process is considered unproven.

While bench-scale testing is typically performed in the laboratory under controlled conditions, pilot testing is often performed with small-scale skid-mounted systems that allow field testing under typical hydraulic operating conditions. Pilot testing serves to fine-tune the pretreatment scheme, the specific membrane desalination process, and the post-treatment for the planned project at conditions equivalent to those of the full-scale plant. A pilot-scale facility is usually sized for flows that are much smaller than a full-scale facility and may or may not include all post-treatment steps or configurations representative of a full-scale facility (e.g., 4-inch membrane elements typically used at pilot scale versus 8-inch membrane elements used at large full-scale desalination plants). In typical pilot studies, small-scale pilot plants use the same feedwater being considered for the desalination plant; and to ensure that the proposed design will operate properly under seasonal variations in source water quality, it is also important that pilot testing be performed for an entire year. In areas where subsurface intakes are being considered, it is desirable for the pilot plant to use appropriately representative water (e.g., water from a nearby well) because the groundwater will likely have different characteristics than surface waters. For optimizing the desalination membrane process design, parameters such as critical flux and the presence and consequences of viable but nonculturable organisms are determined during the pilot testing period (Winters, 2001; Winters et al., 2007). Pilot plants typically take in minimal volumes of feedwater and recombine their product water and reject stream, so that the discharge is not elevated in salinity; thus, pilot plants themselves do not typically pose threats to the environment.

A larger-scale test facility, also called a demonstration project, can be built to confirm final treated water quality and process reliability. A demonstration-scale facility serves as a larger-scaled, more representative test of the full-scale facility and typically employs configurations and all treatment steps that are planned to be included in the full-scale facility. Due to the larger scale, a demonstration testing could also be used to perform an assessment of environmental impacts or to provide better estimates of treatment costs.

One example of a state-supported pilot and demonstration project program is California's Proposition 50 Initiative. This program's objective is to assist local public agencies with the development of new local potable water supplies through the construction of brackish water and seawater desalination projects and to help advance water desalination technology and its use by means of feasibility studies, research and development, and pilot and demonstration projects (Karajeh, 2006). Of the 48 projects awarded through 2006, 15 are pilot and demonstration projects, representing nearly $17 million of the $46 million that has been allocated.[1]

FINISHED WATER QUALITY CHANGES AND EFFECTS OF EXISTING INFRASTRUCTURE

Significant changes in water quality, such as that experienced when a utility brings a desalination plant or any new water supply source online, can affect the water and wastewater infrastructure and, without proper mitigation steps, can impact the quality of the water delivered to consumers. It is, therefore, important to consider the treatment process and the distribution system as an integrated system and to be mindful of unintended consequences. Thus, when considering any new water supply source including desalination technology, a utility should assess the effects of the finished or blended water quality on existing pipelines as well as wastewater facilities that will receive the desalinated water.

The product water quality issue results from the inherent efficiency of desalination technology. That is, the desalination process removes dissolved minerals, producing permeate with low carbonate alkalinity. The reduction of carbonate alkalinity makes permeate or finished water unstable and prone to wide variations in pH due to low buffering capacity (Seacord, 2006b). Lack of carbonate alkalinity and calcium may also contribute to increased corrosion, because protective calcium carbonate films cannot be deposited on pipe walls. In addition, monovalent ions (e.g., chlorides) and gases (e.g., carbon dioxide) may pass through RO membranes to a larger degree than other ions or molecules and contribute to corrosion potential. Post-treatment processes that adjust alkalinity and pH should be incorporated in a desalination plant design to stabilize the permeate water quality, thus minimizing the potential for corrosion of the transmission and distribution system pipeline. Recent research has been directed at this issue and, as a result, general post-treatment strategies have been developed for preventing adverse effects on system infrastruc-

[1] See http://www.owue.water.ca.gov/recycle/DesalPSP/FinalFundingAwards2006.pdf and http://www.owue.water.ca.gov/recycle/DesalPSP/FinalFundingAwards2005.pdf.

ture and water quality (see Box 7-1). However, project-specific water quality and blending analyses should be performed to develop adequate post-treatment systems because all water pipeline systems are unique.

The seawater desalination plant at Ashkelon, Israel, serves as an example of the need for attention to the effects of water quality changes on downstream wastewater systems. At the Ashkelon plant, low alkalinity in the desalinated product water created problems in the wastewater treatment plant. Specifically, sufficient alkalinity is needed to buffer the pH during the nitrification and denitrification process, or the pH will drop and negatively impact the autotrophic nitrifying bacteria, which are pH sensitive. Ashkelon operators now manage the alkalinity at levels (150 to 250 mg/L) to not upset the wastewater treatment process (Lahav and Birnac, 2007). Consideration should also be given to the effects of blending different sources of supply and what impact various blends could have on the water supply system. As discussed in Chapter 5, membrane-desalinated seawater containing bromide can, upon blending with other treated source waters, react with chlorine disinfectants to form bromine and brominated disinfection by-products. Bromide can also adversely affect the stability of chloramine in finished waters (Richardson et al., 1999). Nevertheless, if water providers understand the potential product water quality issues, they can implement post-treatment technologies to minimize public health and infrastructure risk.

CAPITAL AND OPERATING COSTS

Cost is one of the primary considerations for water providers in selecting new supply sources. New technologies can directly influence cost concerns because there is often uncertainty regarding their capital, operating, and maintenance costs. Given the conservative nature of drinking water supply utilities, experience is a key selection criterion for utilities in choosing engineers, contractors, and material suppliers.

To determine the actual cost of desalination, a facility will need to operate at least as long as the major equipment renewal and replacement cycles. For example, in the case of an RO plant, it would be desirable to operate over a period of the expected average membrane life to demonstrate that the pretreatment processes are properly protecting the membranes and that the expected membrane replacement frequency is realized. Membrane life is generally expected to be between 5 and 7

BOX 7-1
Effects of Blending on Distribution System Water Quality

An American Water Works Association Research Foundation study (Taylor et al., 2005) evaluated the impacts of blending desalinated water with treated ground- and surface waters. Using a pilot-scale facility and actual distribution system piping, the corrosion potential of various blends of treated surface water, groundwater, and seawater on extracted pipeline materials found in the Tampa Bay, Florida, area were investigated. Surface water was treated by coagulation, sedimentation, ozonation, biological filtration, and, in some cases, using nanofiltration. Seawater was treated using high-pressure reverse osmosis, and groundwater was treated using aeration and softening in some cases. Eighteen pilot distribution systems were developed using various pipeline materials (PVC, galvanized steel, cement lined pipe, ductile iron) with established biofilm extracted from Tampa Bay Water distribution systems. These systems were installed at the research facility to simulate actual field conditions.

This research addressed a variety of water quality topics including iron, copper, and lead release; maintaining chlorine residual; biostability; nitrification; and a comparison of free chlorine and chloramines as a disinfectant residual. The researchers identified specific water quality parameters that were key to the successful blending of these varied supplies. With caution paid in particular to sulfates, chlorides, and alkalinity, these supplies could be blended and distributed to the customer with little to no adverse consequences in the pipelines.

In this 5-year study, researchers found that the product water chemistry controlled the corrosion effects. Alkalinity was the most significant controlling factor to prevent iron release and complaints of red water. Alkalinity below 40 mg/L was found to cause excess metal release in unlined metal or galvanized pipelines. The research further revealed that pipe material has a more significant impact than water source on the longevity of the disinfectant residual. Cement-lined metal pipes and PVC pipes were found to support chloramine residual stability far more effectively than unlined metal and galvanized pipes. Nitrification is a biological process where excess ammonia from the chloramination process becomes available as a food source for bacteria. The likelihood that nitrification will occur is independent of the water sources and occurred only under conditions of diminished disinfectant residual and was promoted by high free ammonia levels. As long as a disinfectant residual greater than 2 mg/L is maintained, microbial communities (e.g., biofilms) should remain stable. It was also determined that unlined metal pipes can result in low disinfectant residual and slightly higher biofilm growth.

Overall, the study revealed that water quality conditions are primarily dependent on the level of treatment (alkalinity and disinfectant) applied to the source water or the pipe material in the distribution system. Therefore, selection of seawater desalination as a component of a utility's water portfolio should not result in adverse water quality conditions as long as proper post-treatment measures are taken (Taylor et al., 2005).

years, with periods of 10 years experienced in some cases.[2] However, there is inevitably some uncertainty associated with the future costs of any water supply project that includes desalination. For example, future increases in power costs remain unknown, although there may be ways tomitigate this concern through long-term power purchase agreements or utilization of renewable energy sources. Changes in environmental regulations, such as California's recent legislation to reduce greenhouse gas emissions by 25 percent by 2020, may also affect the long-term costs of desalination. Alternative project delivery methods, discussed later, can be used to combine all costs into a single tariff to somewhat limit the water provider's risk of exposure to future cost increases.

A challenge exists for utilities that have historically relied on low-cost water supply projects that do not require extensive treatment. This experience has created an expectation within the public that drinking water can be delivered to the tap at unrealistically low costs. Public acceptance of the development of higher-cost projects, such as desalination or water reuse, can be a slow process that comes about only as water supply shortages become more evident. It should be noted that higher drinking water costs have the side benefit of promoting more conservation as the public will be more selective with water supply use in order to minimize cost (Whitcomb, 2005).

PROJECT DELIVERY METHODS

The process of planning, designing, financing, constructing, and operating a desalination facility can be accomplished through a number of different approaches involving the water supplier (typically a public water provider) and multiple private service providers. The concept of bundling project components under a single contractual relationship with an owner has come to be known as a public-private partnership (PPP). Ultimately, the project delivery method affects the amount of risk carried by the public water provider, and it can influence the access to innovative technology. Alternative project delivery methods can offer advantages over the traditional (design-build-bid) model such as reduced total project costs over the life of the project and shorter time to project completion. A review of the contractual framework of the traditional model and several of the most common alternative project delivery

[2] In the case of the Diablo Canyon Power Plant in Avila Beach, California, the water treatment plant uses seawater RO to provide high-quality boiler feedwater. The pretreatment process includes primary (dual-media filters) followed by secondary (multimedia filters) and ultraviolet treatment. The original RO membranes in this facility were not replaced or cleaned between 1992, when operation began, and 2002 (Prato et al., 2002).

methods are provided in Table 7-1, and the advantages and disadvantages for each of these project delivery methods are highlighted as follows.

Traditional Method: Design-Bid-Build

The traditional public project delivery model, referred to as design-bid-build (DBB), allows for a high degree of involvement and control by the public water provider, because the public water provider oversees the design and construction through separate contractual relationships (see Table 7-1). The public water provider is responsible for obtaining all permits, arranging funding, and will own and operate the plant when construction is complete. The DBB approach tends to be well understood by all parties, and the phased approach to project implementation increases transparency and facilitates public review of the contract process.

Potential disadvantages of this approach are that the public water provider bears responsibility for most of the cost, performance, and risk. Additionally, because of sequential project phasing, it usually takes longer to implement DBB projects than other delivery models, and they are vulnerable to further delays if disputes arise among participants.

Most DBB contracts are evaluated and awarded based solely on capital costs with the objective being to obtain the lowest possible construction cost. As a result, DBB projects tend to avoid the use of proprietary processes or equipment, resulting in low-technology solutions. For most desalination projects, operating costs (including energy) are usually larger than capital costs, including interest (see Figure 6-6). Within the DBB approach, neither the contract engineer nor the contractor has an incentive to promote innovative technologies that may increase capital costs while reducing total project costs.

In the DBB model, the capital investment is most frequently financed by a combination of public equity, public indebtedness, and cost sharing from other governmental entities. Long maturity, low interest rates, and the security of investing in a government entity or project, often combined with local or state tax-free status, has made municipal bonds the main method of financing water infrastructure at the local level (Pankratz and Tonner, 2003).

Alternative Project Delivery Methods

Alternative project delivery methods offer several advantages over the traditional model, because the designer, construction contractor, and operator (if included) can provide input into the different stages of project development. For example, the construction contractor could review the construction design in advance and an operator could offer suggestions to the designer that may reduce operating costs and total project costs. Alternative delivery methods can also save time. For example, a delivery method that allows the engineer and contractor to work together may allow staged construction work to begin prior to completion of the entire design as project permits allow. This can enable site work, such as clearing and grubbing, and some structural foundation work to commence earlier than a DBB process may allow.

Three of the most common alternative project delivery methods—design-build (DB), design-build-operate (DBO), and design-build-own-operate-transfer (DBOOT)—are discussed in this section. There are numerous other project delivery methods available as well as variations on those discussed here, and details on these other methods can be found elsewhere (e.g., Beck, 2002; Pankratz and Tonner, 2003).

Design-Build

A DB delivery approach is characterized by a single contractual relationship between the public water provider and a contractor, who develops the project design and oversees its construction (see Table 7-1). This arrangement reduces the potential for conflicts or disputes, thus reducing the potential for delays, while offering single-point accountability. A DB approach will provide the public water provider with a guaranteed cost, schedule, and performance for the project while transferring the resultant risk to the DB contractor. In the DB approach, the public agency may benefit from newer, innovative technology because the contractor is more focused on facility performance rather than on equipment or construction specifications. However, a public water provider must concede some control over design details. Financing of capital cost using this delivery model is the same as that for the DBB model and is discussed further in the section on public and municipal financing.

Design-Build-Operate

A DBO project involves a single contractor for design, construction, and operation (Tables 7-1a and 7-1b). The DBO model streamlines the project schedule and reduces costs by eliminating separate selection processes for engineering, construction, procurement, and operating services. The contractor provides the public water provider with cost, schedule, and performance guarantees assuring that the project will perform as required, and that the equipment will be maintained, repaired, and replaced according to reasonable and measurable standards. Thus, like the DB model, the DBO model approach transfers certain risks from the public water provider to the private contractors. DBO approaches are often used where project performance and the value of the service to

TABLE 7-1a. Summary of Common Project Delivery Methods.

	Project Delivery Model Structure	Description
Design-Bid-Build (DBB)	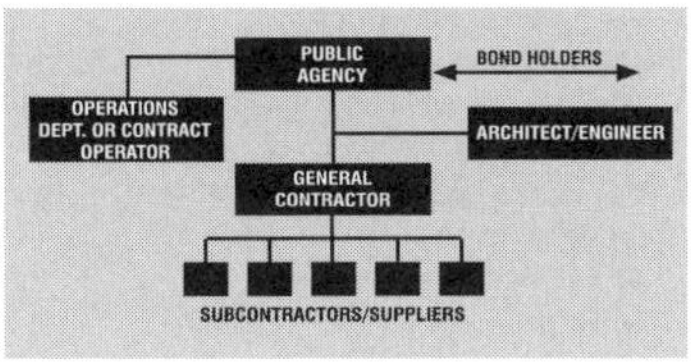	• Owner selects engineer who helps define project, develop bid documents, evaluate bids • Construction awarded to lowest responsive bidder • Construction monitored by engineer or construction manager • Operations by owner or contract operator
Design-Build (DB)	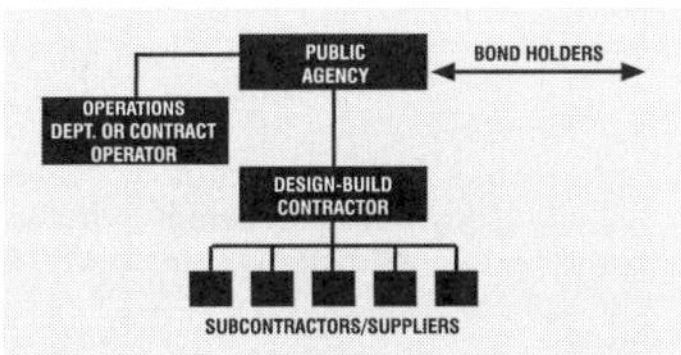	• Owner hires design-build team • Operation by owner or contract operator
Design-Build-Operate (DBO)	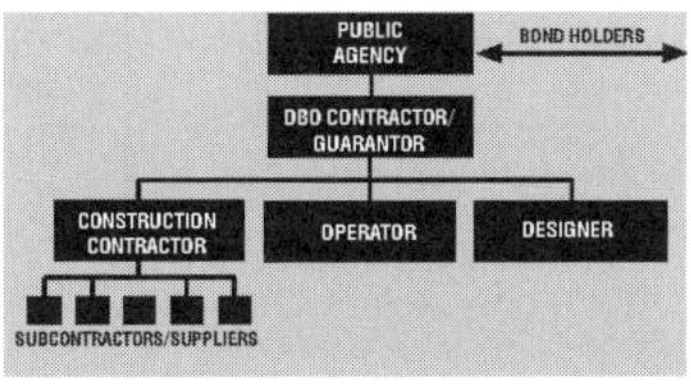	• Involves a single umbrella contractor for overall design, construction, and long-term operation • Owner has wide discretion in how prescriptive or performance-based the process is, but must define existing conditions, inputs (i.e., raw water quality and flows), and expected outcomes
Design-Build-Own-Operate-Transfer (DBOOT)	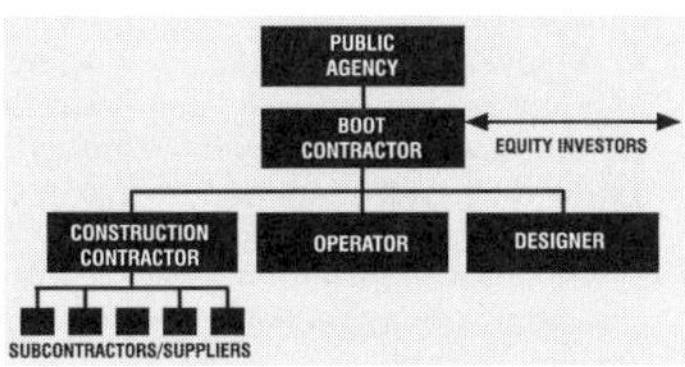	• Similar to DBO except private financing and ownership • Ownership may be transferred to public agency at end of contract term. Contract sets method for valuing facility at that time

SOURCE:: Adapted from Beck (2002).

TABLE 7-1b. Advantages and Disadvantages of Common Project Delivery Methods

	Advantages	Disadvantages	Works best when…
Design-Bid-Build (DBB)	• Well understood by all involved parties • Potential for high degree of control and involvement by owner • Independent oversight of construction contractor	• Segments design, construction, and operation and reduces collaboration • Linear process increases schedule duration • Prone to disputes and creates opportunities for risk avoidance by the designer and construction contractor • Low-bid contactor selection reduces creativity and increases risks of performance problems • Risks are mostly borne by the owner • May not allow for economies of scale in operations • For new technologies, operability may not be the primary design concern	• Operation of facility is minimal or well understood by owner • Project requires a high degree of public oversight • Owner wants to be extensively involved in the design • Schedule is not a priority
Design-Build (DB)	• Collaboration between designer and contractor • Parallel processes reduce duration • Reduces design costs • Reduces potential for disputes between designer and construction contractor • Single point of accountability • Can promote design innovation • Provides more certainty about costs at an earlier stage • Allows owner to assign certain risks to DB team	• Owner may not be as familiar with DB process or contract terms • Reduces owner control and oversight. Owner's rejection of the design, if not based clearly on rights in the contract, can entail large change orders and delay claims • Design and "as-built" drawings not as detailed • Eliminates "independent oversight" role of the designer • Does not inherently include incentives for operability and construction quality as does a DBO or BOOT approach • Higher cost to compete	• Time is critical BUT existing conditions and desired outcomes are well defined • Project uses conventional, well-understood technology • Owner willing to relinquish control over design details • Operational or aesthetic issues are easily defined • Early contractor input will likely save time or money
Design-Build-Operate (DBO)	• Allows designer, construction contractor, and operator to work together collaboratively • Parallel processes reduce duration • Operator input on new technologies and design saves money • DBO contractor has a built-in incentive to assure quality since they will be the long-term operator • Single point of accountability	• Reduces owner involvement • Owner may not be familiar with DBO contracting • High cost to compete may limit competition • Depending on contract terms, may give operator incentives to overcharge for ongoing renewals and replacements or to neglect maintenance near the end of the contract term • Operations contract may limit long-term flexibility • Requires multiphase	• Owner's staff does not have experience operating the type of facility • Input conditions to the facility can be well defined and the number of external influences affecting plant operations are limited • Owner is comfortable with less direct control during design, construction, and operation

continued

	Advantages	Disadvantages	Works best when…
Design-Build-Operate (DBO) *(continued)*	• Allows owner to assign certain risks to DBO contractor • Economies of scale for operations • Collaboration, long-term contract, and appropriate risk allocation can substantially cut costs • Defines long-term expenses for rate setting	contact	
Design-Build-Own-Operate-Transfer (DBOOT)	• Same as DBO and: • Can be used where project expenditures would exceed public borrowing capacity • Beneficial when preserving public credit for other projects is important (i.e. no debt on balance sheet) • Can isolate owner from project risk	• Same as DBO • BUT lack of public financing increases the cost of money	• Public financing cannot be obtained • Transfer of technology risk is important

SOURCE: Adapted from Beck (2002).

be provided is more important than the details of what happens with the various procurement steps along the way.

DBOs are particularly popular with fast-track projects and complex projects that include relatively new technology or specialized operations and maintenance (O&M) expertise (see Box 7-2). With a vested interest in controlling operating expenses, DBO contractors have a greater tendency to accept the risk of employing new and innovative solutions to lower production costs and improve operability. These projects often are driven by the magnitude of total project costs because a single entity is responsible for design, construction, and O&M. Financing of capital cost using this delivery model is the same as that for the DBB model and is discussed further in the section on public and municipal financing.

Design-Build-Own-Operate-Transfer

DBOOT projects are an expansion of the DBO concept in which the contractor also finances the project and initially owns the facility (see Table 7-1a). The public water provider commits to purchase some quantity of water from the desalination facility at an agreed-upon price over some period of time. This water purchase agreement serves as collateral

BOX 7-2
Summary of Project Delivery Method History of the Tampa Bay Seawater Desalination Project

The Tampa Bay Seawater Desalination Project began as a privately owned DBOOT procured project. This delivery method was selected in 1998 by Tampa Bay Water when it was determined that the regulatory and technological risk could be best managed by the private sector. A seawater desalination plant of this size (25 million gallons per day) had never been permitted in the United States, and the technology of surface water desalination had not been successfully implemented in the United States. Tampa Bay Water's initial plan was to structure the deal with the developer that would facilitate an election to purchase the developer's interest in the project at any time, with the intent to purchase the facility after the first replacement cycle of RO membranes had occurred (approximately 5 to 7 years into operation). By that time, the true cost of operating and maintaining the plant would be known, and the plant's performance proven.

Tampa Bay Water was under contractual requirements to build and bring online new alternative drought-resistant supplies in 4 years. Because the initial NPDES permit was taking much longer to issue than originally anticipated, partial interim construction financing was secured. The initial long lead equipment was ordered and the beginning stages of construction were started in order to allow for the facility to be built within the designated schedule. A year later, the developer team failed to secure the appropriate performance bonds necessary to secure the permanent financing. After a 3-month effort to reconfigure the financing, a developer team member (corporate owner) filed for bankruptcy, further complicating the ability of the project team to secure permanent financing and raising the potential cost of private financing. Consequently, Tampa Bay Water took ownership of the facility by buying out the developer's interest in the project earlier than expected with the project near 50 percent completion. Through the buyout provision of the DBOOT contract, Tampa Bay Water assumed the original developer's DBO contracts and contractors for construction completion and plant operation and maintenance. This process transitioned a DBOOT into a DBO arrangement which shifted project ownership and performance risk from the original developer to Tampa Bay Water. In 2003, the DBO contractor subsequently failed the acceptance test by not meeting contractual performance requirements, entered bankruptcy, and did not complete the project. Tampa Bay Water was ultimately able to settle with the contractor in federal bankruptcy court, which allowed pursuit of a replacement contractor for the needed remediation work.

Tampa Bay Water began replacement contractor selection in late 2003 to complete the construction and correct the specific known problems at the desalination plant. In November 2004, Tampa Bay Water retained a new construction completion and operations and maintenance contractor (American Water—Pridesa) using the DBO project delivery method to complete $30 million of repairs and modifications to the facility that would ultimately deliver a reliable desalination water source for the region.

Lessons learned from this experience include the following:

- Contract documents should be created at the beginning of the procurement

continued

process so the developer teams and contractors are submitting proposals for similar contract requirements. Any suggested contract changes should be required to be submitted with the proposals.

- If a DBOOT method is selected, anticipate ownership transfer at any stage of the project.
- Careful consideration should be taken prior to making a decision to transition project ownership in a DBOOT. The assuming owner should understand that they are stepping into the role of the original developer and, therefore, assuming liability for the original developer's decisions.
- A structured and transparent pilot testing program of proposed technologies that supports the design should be conducted prior to selecting a proposal. The pilot program should include pretreatment (including security filters) and RO processes.
- Specific desalination project experience should be a qualification requirement before a proposal is accepted.
- Right of way and property acquisition should be controlled by the public utility because private developers do not have eminent domain authority.

for the contractor to secure private financing for the project. DBOOT contracts contain provisions to transfer ownership of the facility to the public water provider at a mutually agreeable date.

The primary benefit of a DBOOT project delivery is that a private enterprise assumes the technical risk and commercial risk, including the risk of development, permitting, and financing. The public water provider and their ratepayers are relieved of the financial burden of the project and are well insulated from its liabilities and risks; they pay only for water they have contracted to purchase as it is required. The public water provider will be financially protected by performance bonds, professional liability insurances, and liquidated damages provided by the contractor. Although these measures may realize some level of financial protection, certain consequences of plant failure will remain with the public water provider, which is obligated to meet customer demand. Proper prequalification of contractors and prudent review of their capabilities (both technical and financial) and of the proposed plans go far to mitigating these risks.

The Tampa Bay Seawater Desalination Project is an example of a public water provider using alternative delivery methods to implement a desalination project. This project delivery started as a DBOOT and is now being implemented through a DBO arrangement. The history of the delivery method changes and circumstances surrounding those decisions are included in Box 7-2.

A variant of the DBOOT approach is the design-build-own-operate (DBOO) model, where there is no asset transfer at the end of the contract term. Many small DBOO projects have been established to provide desalinated water for hotels and resorts, industries, and small public water providers.

One drawback of both the DBOO and the DBOOT approaches is that lower-interest-rate and tax-exempt public financing is not typically available to private-sector developers in the United States. This may prevent private developers from offering a competitive financing advantage, unless the promise of future public ownership through a DBOOT model can be used to obtain government financing rates. Nevertheless, private financing of desalination projects is on the rise. Prior to 2000, fewer than 5 percent of desalination projects involved private-sector financing. Today, of more than 150 desalination facilities over 5,000 m^3/day that are being planned worldwide (and which identified a source of finance), approximately 51 percent involve private financing (GWI, 2006a). Private financing approaches are discussed further in the next section. Smaller utilities may provide greater opportunity for use of the DBOO or DBOOT process since limited financial resources may preclude cost-effective project financing.

PROJECT FINANCING

The capital cost of desalination projects, similar to other water supply projects, is generally considered to include "up-front" costs such as administrative, design, permitting, property acquisition, and construction costs. These are the costs required to complete or ready a facility for transition into long-term operations. Since these costs are significant, water providers are often required to finance or borrow the funds to implement the project. It is common for the capital cost to be amortized, similar to a home loan, over periods of 20 to 30 years through municipal bonds or other common financing instruments. The water rates established by the water provider will include the amortized capital cost plus operating and maintenance costs and profit for private development projects. Following is a discussion of the public and private ownership financing considerations for desalination projects.

Public and Municipal Financing

The ability for public water providers to obtain financing through bond issues is largely controlled by credit ratings established by three major Wall Street rating agencies (Fitch Ratings, Moody's Investors Service, and Standard & Poor's). Large municipal public water provider and public sewer utilities in the United States have traditionally held high credit ratings. Large public water providers are considered a very stable industry because there have been almost no instances where a utility failed to pay off its debt. Public water providers are advantaged by maintaining a controlled customer base with no competition for service. This customer base provides ample assurance to rating agencies that the customer base will be sustained to repay the debt. For these reasons, public water providers usually receive a general average investment rating in the "A" category ranging from a single "A" to "AAA," a good investment grade rating.

Investment grade ratings are a public water provider's indicator of its ability to borrow funds to finance its capital infrastructure needs and borrow those funds at advantageous interest rates. One factor in a public water provider's rating is the strength of its management practices. These practices include the ability to meet projected agency goals, the flexibility to change with currently changing needs of the public water provider, the continuity of the management team, and the sophistication of its planning process to prepare for meeting future needs (FitchRatings, 2007).

A public water provider will typically issue tax-exempt municipal revenue bonds to build a treatment facility and necessary infrastructure. The benefit of issuing tax-exempt debt is that the bond issuer pays a lower interest rate to the bond buyer than a comparable taxable investment because the bond buyer keeps more of the interest earnings. The revenues from the sale of water from the new treatment facility will typically be pledged to repay the debt. Large public water providers may have the opportunity to issue system debt, which can be repaid from revenues of the entire system, not just the new facility being built. If the borrowing is backed by the system, there is a better opportunity for successful repayment and usually a better credit rating than debt financed solely on the revenues generated by the facility (more typical of small water providers and private water providers). Although there are risks that will be evaluated outside of the project itself, such as extreme water events, most of the credit strength will be reviewed based on the strength of the project itself. Project-level risks such as the contractual foundation; technology, construction, and operations; competitive market exposure; legal structure; counterparty exposure; and financial strength will be evaluated (Box 7-3; Standard & Poor, 2006). Public water providers can also provide additional assurances that the water provider will assist in resolving any issues encountered. In summary, financing desalination projects should not present a significant implementation challenge for large public water providers.

Small public water providers, however, may find it more difficult to finance desalination projects given the limited financial resources in rural areas, the limited access to capital markets, the limited managerial re-

BOX 7-3
Variables Considered in Evaluating Risk for Project Financing

In an evaluation of project risk, the following variables are considered in relation to potential impacts on the operation of the facility and subsequent ability to repay the debt service (Standard & Poor, 2006):

1. **Contractual foundation** is an analysis of the protections provided within the project delivery agreements and an evaluation of how the agreements address the operating risks. The agreements are typically reviewed considering the adequacy and strength of the technology, counterparty credit risk, market risks, and other project characteristics.

2. **Technology, construction, and operations** involve an assessment of the dependability of the design. If the project fails to reach completion or operate as it was contracted then there may be a problem with the ability to meet the debt service requirements. A technical evaluation of the project design by an independent engineer is usually required.

3. **Competitive market exposure** is a comparison of the proposed product price with competing market products. If the market has a natural monopoly, it still must be economically viable with the expected regional costs for the consumer.

4. **Legal structure** involves a review of the entities created to develop this project and an evaluation of the potential impacts to the project should an insolvency of any of the critical players take place.

5. **Counterparty exposure**: a large portion of the project's strength is frequently dependent on other parties such as providers of raw materials, purchases of the project's product, and EPC contractors. The counterparties for the agreements will be evaluated to ascertain the potential risk to the project should the counterparties experience a failure.

6. **Financial strength**: there are many factors that can impact the financial strength of a project such as interest rates, inflation, liquidity and funding. It is also important how the debt is structured and amortized. For example, large balloon payments have proven to be problematic in many corporate and public financing plans.

sources, and the potentially costly concentrate discharge systems. Typical funding alternatives for small systems include state and federal grant and loan programs, conventional commercial loans, and long-term debt-financing mechanisms such as municipal, general obligation, rate revenue, or assessment bonds (NRC, 1997). The Drinking Water State Revolving Fund provides states with the means to establish a revolving fund to provide low-cost loans to public water systems. Desalination plants are eligible for funding through the state revolving funds. Allotments to states for 2006 through 2009 are based on a 2003 needs survey. Each state receives no less than 1 percent of the funding in any fiscal year. In 2007 this resulted in a minimum state allotment of just over $8 million per state (EPA, 2007c). Other options for smaller water facilities include capital facility charges paid by new users as they connect to a system and developer extension policies in which a developer pays the utility to finance a new system or directly bears the cost of the new infrastructure (NRC, 1997).

Private Financing

Desalination projects may utilize traditional equipment financing schemes with 5- to 7-year repayment and prevailing load interest rates, with loan guarantees provided by the project developer. If the project is large enough, development financing with longer repayment periods may be possible, with deferred repayment during construction.

A key feature that DBOO and DBOOT projects have in common is the use of project financing. The lenders' source of repayment is cash flow generated by the project itself; there is little or no recourse for the lender to attempt to recover funds from the public agency. However, the public agency may be expected to make certain guarantees, such as an agreement to purchase the product water assuming contractual commitments are met. Nonrecourse[3] project financing has been used for the world's largest desalination DBOOT project—a 274,000 m^3/day facility at Ashkelon, Israel—as well as for many smaller plants. DBOOT contract durations of up to 30 years are in operation. However, the projects are funded by mixes of equity, bonds, and loans, the mix of which vary for each project. Equity participation is usually a requirement for the main plant designers, the operators, and the developer. Loan terms do not usually match the full term of the DBOOT contract, but terms of up to 18 years have been achieved. Financing has been successfully achieved in both developing nations (such as Trinidad, which has a 30-year design-build-operate-transfer delivery model for the largest seawater RO facility in the Western Hemisphere) and well-established economies (such as Singapore, which as the largest seawater RO DBOOT in Asia). Typical debt-to-equity ratios for a nonrecourse financed desalination project is around 3:1. By ensuring financial risk is contained within the project company, the public agency is unburdened of the risks of developing a new water supply source.

Several different types of bonds may also be available to support private financing. The type of bond depends on the type of entities engaged in the project and their relationships. Corporate bonds, with no tax benefit to investors, are required for entirely private-sector contracts. Bonds exempt from federal and perhaps state and local taxes are possible in

---

[3] A loan where the lending bank is only entitled to repayment from the profits of the project being funded by the loan, not from other assets of the borrower.

DBOO and DBOOT models if the entity purchasing the water is a local government. If the public agency purchasing the water has an equity stake in the project development, presumably municipal bonds could be utilized. If the public agency does not have an equity stake in the project development and more than 10 percent of the funds raised by the bond are utilized for a privately funded project, then municipal private activity bonds presumably would be used.

A number of infrastructure funds, including some based in the United States, have recently increased investment in water assets as a complement to their traditional investments in such privatized infrastructures as toll roads, airports, and railways. To date the infrastructure funds have focused on privatized utilities and a few technology firms but could also be a source of finance for PPP projects.

CONCLUSIONS AND RECOMMENDATIONS

Implementation of desalination projects can be achieved with proper planning and resource availability. Key resources include capital funds and financing capability, funds for electricity or the availability of other sources of energy, access to acceptable source waters, and cost-effective and environmentally sustainable concentrate management options. The implementation process includes the following main components: public involvement programs, regulatory requirements, procurement and project delivery establishment, financing, technology selection, and finished water quality management. Specific recommendations and conclusions follow that pertain to desalination implementation for water providers.

To build trust and educate the public on the key project issues, water providers considering desalination should engage the public as early as possible. Key issues should include environmental impacts, site selection, cost, product water quality, and reliability of the technology. The water supplier can improve public relations by establishing a process to make information readily available to the public including public meetings, newsletters, and websites.

Water providers should meet with regulatory agencies as soon as possible in the planning and development process to begin understanding applicable regulatory requirements. The permitting process involved in the planning, design, or construction of a desalination facility requires significant time and cost. This process can be very cumbersome and can pose implementation challenges to small public water providers. Although the types and complexity of permits vary with project location, concentrate management method, and other specific factors, typically multiple permits are required from federal, state, and local agencies. Thus, it is recommended that the permitting process for a planned desalination facility should begin at the earliest planning stage.

Water providers should consider all project delivery models available and select the most appropriate model. The project delivery method selected can affect the amount of risk carried by the water provider, the access to innovative technology, and the time to project completion. Once a delivery method is selected, contracts should be developed during the early stages of the procurement process so risk allocation is clearly defined prior to the submittal of cost proposals. Contract documents for DB and DBB focus on equipment specifications whereas those for DBO and DBOOT focus on performance (i.e., quality and quantity). A development contractor should be selected based on a balance of financial and technical criteria, as required for the selected project delivery model, including related project experience. Local and state regulations may force water suppliers to consider only certain project delivery models (e.g., DBB).

To evaluate and optimize the process design early in the process, water providers should conduct pilot testing. The pilot process should include all of the key project processes from pretreatment through desalination to evaluate individual component performance as part of the overall system. Pilot processes should be designed to simulate full-scale operations including key parameters such as chemical feed rates, chemical contact times, and filter load rates.

To prevent unanticipated water quality changes, the effects of blending the desalinated product water with other existing sources should be considered during project design. Proven post-treatment technology is available to mitigate potential problems that could otherwise be experienced in the product water distribution systems and downstream wastewater systems.

Desalination water providers should consider potential decommissioning of once-through cooling power plants and wastewater plants when evaluating co-location benefits. Power plant cooling water and wastewater plant effluent can serve as a means of diluting concentrate prior to discharge to the environment and can significantly reduce capital costs associated with intake and outfall construction. However, should these plants be decommissioned, alternative methods of concentrate management will have to be developed. Potential concentrate management options might include offshore multiple diffuser discharges and deep-well injection, among others.

8

A Strategic Research Agenda for Desalination

As noted in Chapter 3, desalination is likely to have a niche in the water management portfolio of the future, although the significance of this niche cannot be definitively determined at this time. The potential for desalination to meet anticipated water demands in the United States is not constrained by the source water resources or the capabilities of current technology, but instead it is constrained by financial, social, and environmental factors. Over the past 50 years the state of desalination technology has advanced substantially, and improvements in energy recovery and declining membrane material costs have made brackish water and seawater desalination a more reasonable option for some communities. However, desalination remains a higher-cost alternative for water supply in many communities, and concerns about potential environmental impacts continue to limit the application of desalination technology in the United States. For inland desalination facilities, there are few, if any, cost-effective environmentally sustainable concentrate management technologies. Meanwhile, as noted in Chapter 2, there is no integrated and strategic direction to current federal desalination research and development efforts to help address these concerns.

In this chapter, long-term research goals are outlined for advancing desalination technology and improving the ability of desalination to address U.S. water supply needs. A strategic national research agenda is then presented to address these goals. This research agenda is broadly conceived and includes research that could be appropriately funded and conducted in either the public or private sectors. The committee recognizes that research cannot address all barriers to increased application of desalination technology in regions facing water scarcity concerns; therefore, practical implementation issues are discussed separately in Chapter 7. Recommendations related to implementing the proposed research agenda are also provided in this chapter.

LONG-TERM RESEARCH GOALS

Based on the committee's analyses of the state of desalination technology, potential environmental impacts, desalination costs, and implementation issues in the United States (see Chapters 4-7), the committee developed two overarching long-term goals for further research in desalination:

1. Understand the environmental impacts of desalination and develop approaches to minimize these impacts relative to other water supply alternatives, and

2. Develop approaches to lower the financial costs of desalination so that it is an attractive option relative to other alternatives in locations where traditional sources of water are inadequate.

Understanding the potential environmental impacts of desalination in both inland and coastal communities and developing approaches to mitigate these impacts relative to other alternatives are essential to the future of desalination in the United States. The environmental impacts of both source water intakes and concentrate discharge remain poorly understood. Although the impacts of coastal desalination are suspected to be less than those of other water supply alternatives, the uncertainty about potential site-specific impacts and their mitigation are large barriers to the application of coastal desalination in the United States. This uncertainty leads to stakeholder disagreements and a lengthy and costly planning and permitting process. For inland desalination, uncertainties remain about the sustainability of brackish groundwater resources and the environmental impacts from concentrate discharge to surface waters. Without rigorous scientific research to identify specific potential environmental impacts (or a lack of impacts), planners cannot assess the feasibility of desalination at a site or determine what additional mitigation steps are needed. Once potential impacts are clearly understood, research can be focused on developing approaches to minimize these impacts.

The second goal focuses on the cost of desalination relative to the cost of other water supply alternatives. At present, costs are already low enough to make desalination an attractive option for some communities, especially where concentrate management costs are modest. In fact, desalination plants are being studied or implemented in at least 30 municipalities nationwide (GWI, 2007). The economic costs of desalination, however, as well as the costs of water supply alternatives, are locally variable. Costs are influenced by factors such as source water quality, siting considerations, potential environmental impacts, local regulations and permitting requirements, and available concentrate management options. Desalination remains a higher-cost alternative for many locations, and increasing awareness of potential environmental impacts is raising the costs of permitting and intake and outfall configurations in the United States. Inland communities considering brackish groundwater desalination may soon face more restrictions on surface water discharge and, therefore, will have fewer low-cost alternatives for concentrate management. Meanwhile, the future costs of energy are uncertain. If the total costs of desalination (including environmental costs) were reduced relative to other alternatives, desalination technology would become an attractive alternative to help address local water supply needs.

STRATEGIC DESALINATION RESEARCH AGENDA

The committee identified research topics as part of a strategic agenda to address the two long-term research goals articulated earlier. This agenda is driven by determination of what is necessary to make desalination a competitive option among other water supply alternatives. The agenda is broadly conceived, including research topics of clear interest to the public sector—and therefore of interest for federal funding—*and* research that might be most appropriately funded by private industry. The suggested research areas are described in detail below and are summarized in Box 8-1. Specific recommendations on the roles of federal and nonfederal organizations in funding the agenda are described in an upcoming section.

BOX 8-1
Priority Research Areas

The committee has identified priority research areas to help make desalination a competitive option among water supply alternatives for communities facing water shortages. These research areas, which are described in more detail in the body of the chapter, are summarized here. The highest priority topics are shown in bold. Some of this research may be most appropriately supported by the private sector. The research topics for which the federal government should have an interest—where the benefits are widespread and where no private-sector entities are willing to make the investments and assume the risk—are marked with asterisks.

GOAL 1. Understand the environmental impacts of desalination and develop approaches to minimize these impacts relative to other water supply alternatives

1. **Assess environmental impacts of desalination intake and concentrate management approaches****

continued

a. Conduct field studies to assess environmental impacts of brackish groundwater development**
b. Develop protocols and conduct field studies to assess the impacts of concentrate management approaches in inland and coastal settings**
c. Develop laboratory protocols for long-term toxicity testing of whole effluent to assess long-term impacts of concentrate on aquatic life**
d. Assess the environmental fate and bioaccumulation potential of desalination-related contaminants**

2. Develop improved intake methods at coastal facilities to minimize impingement of larger organisms and entrainment of smaller ones**

3. Assess the quantity and distribution of brackish water resources nationwide**

4. Analyze the human health impacts of boron, considering other sources of boron exposure, to expedite water-quality guidance for desalination process design**

GOAL 2. Develop approaches to lower the financial costs of desalination so that it is an attractive option relative to other alternatives in locations where traditional sources of water are inadequate

5. Improve pretreatment for membrane desalination
   a. Develop more robust, cost-effective pretreatment processes
   b. Reduce chemical requirements for pretreatment
6. Improve membrane system performance
   a. Develop high-permeability, fouling-resistant, high-rejection, oxidant-resistant membranes
   b. Optimize membrane system design
   c. Develop lower-cost, corrosion-resistant materials of construction
   d. Develop ion-selective processes for brackish water
   e. Develop hybrid desalination processes to increase recovery
7. Improve existing desalination approaches to reduce primary energy use
   a. Develop improved energy recovery technologies and techniques for desalination
   b. Research configurations and applications for desalination to utilize low-grade or waste heat**
   c. Understand the impact of energy pricing on desalination technology over time**
   d. Investigate approaches for integrating renewable energy with desalination**
8. Develop novel approaches or processes to desalinate water in a way that reduces primary energy use**

GOAL 1 and 2 Crosscuts

9. Develop cost-effective approaches for concentrate management that minimize potential environmental impacts**

Research on Environmental Impacts

The following research topics address Goal 1 to understand the environmental impacts of desalination and develop approaches to minimize those impacts relative to other water supply alternatives.

1. Assess environmental impacts of desalination intake and concentrate management approaches

As discussed in Chapter 5, the environmental impacts of desalination source water intake and concentrate management approaches are not well understood. Source water intakes for coastal desalination can create entrainment concerns with small organisms and impingement issues for larger organisms. For inland groundwater desalination, there are potential concerns regarding overpumping, water quality changes, and subsidence. The possible environmental impacts of concentrate management approaches range from effects on aquatic life in surface water discharges to the contamination of drinking water aquifers in poorly designed injection wells or ponds. Both site-specific studies and broad analyses of relative impacts would help communities weigh the alternatives for meeting water supply needs. The specific research needs are described as follows.

1a. Conduct field studies to assess environmental impacts of seawater intakes. Measurements and modeling of the extent of mortality of aquatic or marine organisms due to impingement and entrainment are needed. There have been numerous studies on such impacts of power plants, and extrapolation of such effects to desalination facilities should be performed.

1b. Conduct field studies to assess environmental impacts of brackish groundwater development. The general environmental interactions between wetlands, freshwater, and brackish aquifers for inland sources have not been documented under likely brackish water development scenarios. While site-specific evaluation of any location will be necessary for developing a brackish water resource, the lack of synthesized information is an impediment to the use of this resource for smaller communities with limited resources.

1c. Develop protocols and conduct field studies to assess the impacts of concentrate management approaches in inland and coastal settings. Comprehensive studies analyzing impacts of concentrate discharge at marine, estuarine, and inland desalination locations are needed.

Adequate site-specific baseline studies on potential biological and ecological effects are necessary prior to the development of desalination facilities because biological communities in different geographic areas will have differential sensitivity, but a comprehensive synthesis would be valuable once several in-depth studies have been conducted. Protocols should be developed to define the baseline and operational monitoring, reference sites, lengths of transects, and sampling frequencies. Planners would benefit from clear guidance on appropriate monitoring and assessment protocols. Environmental data should be collected for at least 1 year in the area of the proposed facility before a desalination plant with surface water concentrate discharge comes online so that sufficient baseline data on the ecosystem are available with which to compare postoperating conditions. Once a plant is in operation, monitoring of the ecological communities (especially the benthic community) receiving the concentrate should be performed periodically for at least 2 years at multiple distances from the outflow pipe and compared to reference sites.

For inland settings, additional regional hydrogeology research is needed on the distribution, thickness, and hydraulic properties of formations that could be used for disposal of concentrate via deep-well injection. Much information is already available about the potential for deep-well injection in states such as Florida and Texas, although suitable geologic conditions may exist in other states as well. Inventories of industrial and commercial brine-disposal wells and producing and abandoned oil fields should be synthesized and used to develop a suitable protocol for further hydrogeological investigations, as appropriate. This research would provide valuable assistance to small communities that typically do not have the resources available to support extensive hydrogeological investigations.

1d. Develop laboratory protocols for long-term toxicity testing of whole effluent to assess long-term impacts of concentrate on aquatic life. Standard acute toxicity tests as defined by the U.S. Environmental Protection Agency (EPA) are generally 96 hours in duration and use larval or juvenile stages of certain fish and invertebrate species with a series of effluent dilutions and a control. The end point is whether the test organisms survive or not. Chronic tests, according to EPA, are typically 7 days in duration when using larval stages of fish and invertebrate species, and the end points of the tests are sublethal, such as growth reduction. Typical chronic toxicity protocols were designed for testing municipal or industrial wastewater treatment plant effluent, which typically contains higher levels of toxic chemicals than the concentrate from desalination plants. To assess the impacts of desalination effluent, a protocol should be developed to analyze the longer-term effects (over whole life cycles) on organisms that live in the vicinity of desalination plants (as opposed to the standard species used in EPA-required toxicity testing). These laboratory-based tests should then be used to examine the impacts of whole effluent (and various dilutions) from different desalination plants on a variety of different taxa at numerous representative sites from key ecological regions.

1e. Assess the environmental fate and bioaccumulation potential of desalination-related contaminants. Desalination concentrate contains more than just salts and may include various chemicals that are used in pretreatment and membrane cleaning, antiscaling and antifoulant additives, and metals that may leach from corrosion. Some of these chemicals (e.g., antifoulants, copper leached from older thermal desalination plants) or chemical by-products (e.g., trihalomethanes produced as a result of pretreatment with chlorine) are likely to bioaccumulate in organisms. Investigations into the loading and environmental fate of desalination-related chemicals should be included in modeling and monitoring programs. The degree to which various chemicals biodegrade or accumulate in sediments should also be investigated. High priority should be given to polymer antiscalants, such as polycarbonic acids and polyphosphate, which may increase primary productivity. Corrosion-related metals and disinfection by-products should also be investigated. In conjunction with the field studies described earlier, representative species, preferably benthic infauna along the transects and from the reference (control) site, should be analyzed for bioaccumulative contaminants. Because little is known about the potential of some other desalination chemicals that can be discharged in concentrate to bioaccumulate (e.g., polyphosphate, polycarbonic acid, polyacrylic acid, polymaleic acid), research should be conducted into their toxicity and bioaccumulation potential.

2 Develop improved intake methods at coastal facilities to minimize impingement of larger organisms and entrainment of smaller ones

Although intake and screen technology is rapidly developing, continued research and development is needed in the area of seawater intakes to develop cost-effective approaches that minimize the impacts of impingement and entrainment for coastal desalination facilities. Current technology development has focused on subsurface intakes and advanced screens or curtains, and these recent developments should be assessed to determine the costs and benefits of the various approaches. Other innovative concepts could also be considered that might deter marine life from entering intakes.

3. Assess the quantity and distribution of brackish water resources nationwide

Sustainable development of inland brackish water resources requires maps and synthesized information on total dissolved solids of the groundwater, types of dominant solutes (e.g., NaCl, $CaSO_4$), thickness, and depth to brackish water. The only national map of brackish water resources available (Feth, 1965; Figure 1-1) simply shows depth to saline water. Newer and better solute chemistry data collected over the past 40 years exist in the files of private, state, and federal offices but are not generally organized for use in brackish water resources investigations. Using the aforementioned information, basin analyses, analogous to the U.S. Geological Survey Regional Aquifer System Analysis program for freshwater, could be developed, emphasizing regions facing near-term water scarcity concerns. These brackish water resource investigations could also be conducted at the state level. The data, once synthesized, could be utilized for desalination planning as well as for other water resources and commercial development scenarios.

4. Analyze the human health impacts of boron, considering other sources of boron exposure, to expedite water-quality guidance for desalination process design

Typical single-pass reverse osmosis (RO) desalination processes do not remove all the boron in seawater; thus, boron can be found at milligram-per-liter levels in the finished water. Boron can be controlled through treatment optimization, but that treatment has an impact on the cost of desalination. A range of water quality levels (0.5 to 1.4 mg/L) have been proposed as protective of public health based on different assumptions in the calculations. Because of the low occurrence of boron in most groundwater and surface water, the EPA has decided not to develop a maximum contaminant level for boron and has encouraged affected states to issue guidance or regulations as appropriate (see Chapter 5). Additional analysis of existing boron toxicity data is needed, considering other possible sources of boron exposure in the United States, to support guidance for desalination process design that will be suitably protective of human health.

Research to Lower the Costs of Desalination

The following research topics address Goal 2 to develop approaches to lower the costs of desalination so that it is an attractive option relative to other alternatives in locations where traditional sources of water are inadequate. As a broadly conceived agenda, some of this research may be most appropriately supported by the private sector. The appropriate roles of governmental and nongovernmental entities to fund the research agenda are discussed later in the chapter.

5. Improve pretreatment for membrane desalination

Pretreatment is necessary to remove potential foulants from the source water, thereby ensuring sustainable operation of the RO membranes at high product water flux and salt rejection. Research to improve the pretreatment process is needed that would develop alternative, cost-effective approaches.

5a. Develop more robust, cost-effective pretreatment processes. Membrane fouling is one of the most problematic issues facing seawater desalination. Forms of fouling common with RO membranes are organic fouling, scaling, colloidal fouling, and biofouling. All forms of fouling are caused by interactions between the foulant and the membrane surface. Improved pretreatment that minimizes these interactions will reduce irreversible membrane fouling. Alteration of solution characteristics can improve the solubility of the foulants, preventing their precipitation or interaction with the membrane surface. Such alteration could be chemical, electrochemical, or physical in nature.

Membranes such as microfiltration (MF) and ultrafiltration (UF) have several advantages over traditional pretreatment (e.g., conventional sand filtration) because they have a smaller footprint, are more efficient in removing smaller foulants, and provide a more stable influent to the RO membranes. Additional potential benefits of MF or UF pretreatment are increased flux, increased recovery, longer membrane life, and decreased cleaning frequency. More research is necessary in order to optimize the pretreatment membranes for more effective removal of foulants to the RO system, to reduce the fouling of the pretreatment membranes, and to improve configuration of the pretreatment membranes to maximize cost reduction.

5b. Reduce chemical requirements for pretreatment. Antiscalants, coagulants, and oxidants (such as chlorine) are common chemicals applied in the pretreatment steps for RO membranes. Although these chemicals are added to reduce fouling, they add to the operational costs, can reduce the operating life of membranes, and have to be disposed of properly or they can adversely impact aquatic life (see Chapter 5). Antiscalants may also enhance biofouling, so alternative formulations or approaches should be examined. Research is needed on alternative formulations or approaches (including membrane pretreatment) to reduce the chemical requirements of the pretreatment process, both to reduce overall cost and to decrease the environmental impacts of desalination.

6. Improve membrane system performance

Sustainable operation of the RO membranes at the designed product water flux and salt rejection is a key to the reduction of desalination process costs. In addition to effective pretreatment, research to optimize the sustained performance of the RO membrane system is needed.

6a. Develop high-permeability, fouling-resistant, high-rejection, oxidant-resistant membranes. New membrane designs could reduce the treatment costs of desalination by improving membrane permeability and salt rejection while increasing resistance to fouling and membrane oxidation. Current membrane research to reduce fouling includes altering the surface charge, increasing hydrophilicity, adding polymers as a barrier to fouling, and decreasing surface roughness.

Oxidant-resistant membranes enable feedwater to maintain an oxidant residual that will reduce membrane fouling due to biological growth. Current state-of-the-art thin-film composite desalination membranes are polyamide based and therefore are vulnerable to damage by chlorine or other oxidants. Thus, when an oxidant such as chlorine is added to reduce biofouling, dechlorination is necessary to prevent structural damage. Additionally, trace concentrations of chlorine may be present in some feedwaters. Cellulose-derivative RO membranes have much higher chlorine tolerance; however, these membranes have a much lower permeability than thin-film composite membranes and operate under a narrower pH range. Therefore, there is a need to increase the oxidant tolerance of the higher-permeability membranes. Lower risk of premature membrane replacement equates to overall lower operating costs.

Past efforts to synthesize RO membranes with high permeability often resulted in reduced rejection and selectivity. There is a need to develop RO membranes with high permeability without sacrificing selectivity or rejection efficiency. Recent research on utilizing nanomaterials, such as carbon nanotubes, as a separation barrier suggest the possibility of obtaining water fluxes much higher than that of traditional polymeric membranes.

The development of membranes that are more resistant to degradation from exposure to cleaning chemicals will extend the useful life of a membrane module. The ability to clean membranes more frequently can also decrease energy usage because membrane fouling results in higher differential pressure loss through the modules. By extending the life of membrane modules, the operating and maintenance cost will be reduced by the associated reduction in membrane replacements required.

6b. Optimize membrane system design. With the development of high-flux membranes and larger-diameter membrane modules, new approaches for optimal RO system design are needed to avoid operation under thermodynamic restriction (see Chapter 4) and to ensure equal distribution of flux between the leading and tail elements of the RO system. The key variables for the system design will involve the choice of optimal pressure, the number of stages, and number and size of membrane elements at each stage. An optimal system configuration may also involve hybrid designs where one type of membrane (e.g., intermediate flux, highly fouling-resistant) is used in the leading elements followed by high-flux membranes in the subsequent elements. Fouling can be mitigated by maintaining high crossflow velocity; thus, fouling-resistant membranes may be better served in the downstream positions where lower crossflow velocity is incurred. Thus, additional engineering research on membrane system design is needed to optimize performance with the objective of reducing costs.

6c. Develop lower-cost, corrosion-resistant materials of construction. The duration of equipment life in a desalination plant directly relates to the total costs of the project. Saline and brackish water plants are considered to be a corrosive environment due to the high levels of salts in the raw water. The development and utilization of corrosion-resistant materials will minimize the frequency of equipment or appurtenance replacement, which can significantly reduce the total project costs.

6d. Develop ion-selective processes for brackish water. Some slightly brackish waters could be made potable simply though specific removal of certain contaminants, such as nitrate or arsenite, while removing other ions such as sodium, chloride, and bicarbonate at a lower rate. High removal rates of all salts are not necessary for such waters. Ion-specific separation processes, such as an ion-selective membrane or a selective ion-exchange resin, should be able to produce potable water at much lower energy costs than those processes that fully desalinate the source water. Ion-selective removal would also create fewer waste materials requiring disposal. Ion-selective processes would be useful for mildly brackish groundwater sources with high levels of nitrate, uranium, radium, or arsenic. Such an ion-selective process could also be used to optimize boron removal following RO desalination of seawater.

6e. Develop hybrid desalination processes to increase recovery. Overall product water recovery in a desalination plant can be increased through the serial application of more than one desalination process. For example, an RO process could be preceded by a "tight" nanofiltration process, allowing the RO to operate at a higher recovery than it could with less aggressive pretreatment. Other options could be devised, including hybrid thermal and membrane processes to increase the overall recovery of the process. As noted in Chapter 4, the possible hybrid combinations of desalination processes are limited only by ingenuity and identification of economically viable applications. Hybridization also offers opportunities for reducing desalination production costs and expanding the flexibility of operations, especially when co-located with power plants, but hybridization also increases plant complexity and raises challenges in operation and automation.

7. Improve existing desalination approaches to reduce primary energy use

Energy is one of the largest annual costs in the desalination process. Thus, research to improve the energy efficiency of desalination technologies could make a significant contribution to reducing costs.

7a. Develop improved energy recovery technologies and techniques for desalination. Membrane desalination is an energy-intensive process compared to treatment of freshwater sources. Modern energy recovery devices operate at up to 96 percent energy recovery (see Chapter 4), although these efficiencies are lower at average operating conditions. The energy recovery method in most common use today is the energy recovery (or Pelton) turbine, which achieves about 87 percent efficiency. Many modern plants still use Pelton wheels because of the higher capital cost of isobaric devices. Thus, opportunities exist to improve recovery of energy from the desalination concentrate over a wide operating range and reduce overall energy costs.

7b. Research configurations and applications for desalination to utilize low-grade or waste heat. Industrial processes that produce waste or low-grade heat may offer opportunities to lower the operating cost of the desalination process if these heat sources are co-located with desalination facilities (see Box 4-8). Low-grade heat can be used as an energy source for desalination via commercially available thermal desalination processes. Hybrid membrane-thermal desalination approaches offer additional operational flexibility and opportunities for water-production cost savings. Research is needed to examine configurations and applications of current technologies to utilize low-grade or waste heat for desalination.

7c. Understand the impact of energy pricing on existing desalination technology over time. Energy is one of the largest components of cost for desalination, and future changes in energy pricing could significantly affect the affordability of desalination. Research is needed to examine to what extent the economic and financial feasibility of desalination may be threatened by the uncertain prospect of energy price increases in the future for typical desalination plants in the United States. This research should also examine the costs and benefits of capital investments in renewable energy sources.

7d. Investigate approaches for integrating renewable energy with desalination. Renewable energy sources could help mitigate future increases in energy costs by providing a means to stabilize energy costs for desalination facilities while also reducing the environmental impacts of water production. Research is needed to optimize the potential for coupling various renewable energy applications with desalination.

8. Develop novel approaches or processes to desalinate water in a way that reduces primary energy use

Because the energy of RO is only twice the minimum energy of desalination, even novel technologies are unlikely to create step change (>25 percent) reductions in absolute energy consumption compared to the best current technology (see, e.g., Appendix A). Instead, substantial reductions in the energy costs of desalination are more likely to come through the development of novel approaches or processes that optimize the use of low-grade heat. Several innovative desalination technologies that are the focus of ongoing research, such as forward osmosis, dewvaporation, and membrane distillation, have the capacity to use low-grade heat as an energy source. Research into the specific incorporation of waste or low-grade heat into these or other innovative processes could greatly reduce the amount of primary energy required for desalination and, thus, overall desalination costs.

Crosscutting Research

Research topics in this category benefit both Goal 1, for environmental impacts, and Goal 2, for lowering the cost of desalination.

9. Develop cost-effective approaches for concentrate management that minimize potential environmental impacts

Research objectives related to concentrate management are crosscutting, because they address both the need to understand and minimize environmental impacts and the need to reduce the total cost of desalination. For coastal concentrate management, research is needed to develop improved diffuser technologies and subsurface injection approaches and to examine their costs and benefits relative to current disposal alternatives.

The high cost of inland concentrate management inhibits inland brackish water desalination. Low- to moderate-cost concentrate management alternatives (i.e., subsurface injection, land application, sewer discharge, and surface water discharge) can be limited by the salinity of the concentrate and by location and climate factors; in some scenarios all of these options may be restricted by site-specific conditions, leaving zero liquid discharge (ZLD) as the only alternative for consideration. ZLD options currently include evaporation ponds and energy-intensive processes, such as brine concentrators or crystallizers, followed by landfilling. These options have high capital or operating costs. Research to improve recovery in the desalination process and thereby minimize the initial volume of concentrate could enhance the practical viability of several concentrate management options for inland desalination. This is particularly true for the concentrate management options that are characterized by high costs per unit volume of the concentrate flow treated and for approaches that are not applicable to large concentrate flows, such as thermal evaporation or evaporation ponds. Advancements are also needed that reduce the capital costs and improve the energy efficiency of thermal evaporation processes. Conventional concentrate management options that involve simple equipment are not likely to see significant cost reductions through additional research.

The reuse of high-salinity concentrates and minerals extracted from them should be further explored and developed to help mitigate environmental impacts while generating revenues that can help offset concentrate management costs. Possibilities include selective precipitation of marketable salts, irrigation of salt-tolerant crops, supplements for animal dietary needs, dust suppressants, stabilizers for road base construction, or manufacture of lightweight fire-proof building materials. Studies are necessary to determine the most feasible uses and to develop ways to prepare the appropriate product for various types of reuse. For all possible uses, site-specific limitations and local and state regulations will need to be considered. Because the transportation costs greatly affect the economics of reuse, a market analysis would also be needed to identify areas in the United States that could reasonably utilize products from desalination concentrate.

Highest Priority Research Topics

All of the topics identified are considered important, although three topics (1, 2, and 9 above) were deemed to be the highest priority research topics: (1) assessing the environmental impacts of desalination intake and concentrate management approaches, (2) developing improved intake methods to minimize impingement and entrainment, and (3) developing cost-effective approaches for concentrate management that minimize environmental impacts. These three research areas are considered the highest priorities because this research can help address the largest barriers (or showstoppers) to more widespread use of desalination in the United States. Uncertainties about potential environmental impacts will need to be resolved and cost-effective mitigation approaches developed if desalination is to be more widely accepted. Research to develop cost-effective approaches for concentrate management is critical to enable more widespread use of desalination technologies for inland communities. As noted in Chapter 4, the cost of concentrate management can double or triple the cost of the desalination for some inland communities.

Research may also reduce the costs of desalination. Any cost improvement will help make desalination an attractive option for communities addressing water shortages. However, the committee does not view these process cost issues as the major limitation to the application of desalination in the United States today.

IMPLEMENTING THE RESEARCH AGENDA

In the previous section, the committee proposed a broad research agenda that, if implemented, should improve the capacity of desalination to meet future water needs in the United States by further examining and addressing its environmental impacts and reducing its costs relative to other water supply alternatives. Implementing this agenda requires federal leadership, but its success depends on participation from a range of entities, including federal, state, and local governments, nonprofit organizations, and the private sector. A strategy for implementing the research agenda is suggested in the following section. This section also includes suggestions for funding the agenda and the appropriate roles of government and nongovernmental entities.

Supporting the Desalination Research Agenda

A federal role is appropriate for research that provides a "public good." Specifically, the federal government should have an interest in funding research where the benefits are widespread but where no private-sector entities are willing to make the investment and assume the risks. Thus, for example, research that results in significant environmental benefits should be in the federal interest because these benefits are shared by the public at large and cannot be fully captured by any entrepreneur. Federal investment is also important where it has "national significance"—where the issues are of large-scale concern; they are more than locally, state-, or regionally specific; and the benefits accrue to a large swath of the public.

Based on the aforementioned criteria, the proposed research agenda contains many topic items that should be in the federal interest (see topics marked with asterisks in Box 8-1). The research topics in support of Goal 1 (see Box 8-1) are directed at environmental issues that are largely "public good" issues. Some of the needed environmental research will, by nature, be site-specific, and purely site-specific research is not of great federal interest. Thus, there is a clear role for state and local agencies to support site-specific research. The federal government, however, should have an interest in partnering with local communities to conduct more extensive field research from which broader conclusions of environmental impacts can be drawn or which would significantly contribute to a broader meta-analysis. This meta-analysis could especially benefit small water supply systems. Also, there should be federal interest in establishing general protocols for field evaluations and chronic bioassays that could then be adapted for site-specific studies.

The research needed to support the attainment of Goal 2 includes several topics that are clearly in the federal interest, as defined earlier. These include efforts to reduce prime energy use, to integrate renewable energy resources within the total energy picture and increase reliance upon them, and to understand the impacts of energy pricing on the future of desalination (see highlighted topics in Box 8-1). However, Goal 2 also includes a number of research topics that may be more appropriately funded by the private sector or nongovernmental organizations, assuming that these entities are willing to assume the risks of the research investment. Indeed, private industry already spends far more on research and development for desalination than the federal government (see Chapter 2) and is already making substantial progress in the improvement of existing membrane performance, developing better pretreatment alternatives, and developing improved energy recovery devices. To avoid duplication and to optimize available research funding, government programs should focus instead on research and development with widespread possible benefits that would otherwise go unfunded because private industry is unwilling to make the investment. Finally, the crosscutting topic to develop cost-effective methods of managing concentrates for inland communities, which impacts Goals 1 and 2, is also in the federal interest.

Federal Research Funding

The optimal level of federal investment in desalination research is inherently a question of public policy. Although the decision should be informed by science, it is not—at its heart—a scientific decision. However, several conclusions emerged from the committee's analysis of current research and development funding (see Chapter 2) that suggest the importance of strategic integration of the research program. The committee concluded that there is no integrated and strategic direction to the federal desalination research and development efforts. Continuation of a federal program of research dominated by congressional earmarks and beset by competition between funding for research and funding for construction will not serve the nation well and will require the expenditure of more funds than necessary to achieve specified goals.

To ensure that future federal investments in desalination research are integrated and prioritized so as to address the two major goals identified in this report, the federal government will need to develop a coordinated strategic plan that utilizes the recommendations of this report as a basis. It is beyond the committee's scope to recommend specific plans for improving coordination among the many federal agencies that support desalination research. Instead, responsibility for developing the plan should rest with the Office of Science and Technology Policy's (OSTP's) National Science and Technology Council (NSTC) because "this Cabinet-level Council is the principal means within the executive branch to coordinate science and technology policy across the diverse entities that make up the Federal research and development enterprise."[1] For example, the NSTC's Subcommittee on Water Availability and Quality has membership representing more than 20 federal agencies and recently released "A Strategy for Federal Science and Technology to Support Water Avail-

[1] For more information, see http://www.ostp.gov/nstc/index.html.

ability and Quality in the United States" (SWAQ, 2007). Representatives of the National Science Foundation, the Bureau of Reclamation, the Environmental Protection Agency, the National Oceanographic and Atmospheric Administration, the Office of Naval Research, and the Department of Energy should participate fully in the development of the strategic federal plan for desalination research and development. Five years into the implementation of this plan, the OSTP should evaluate the status of the plan, whether goals have been met, and the need for further funding.

A coordinated strategic plan governing desalination research at the federal level along with effective implementation of the research plan will be the major determinants of federal research productivity in this endeavor. The committee cannot emphasize strongly enough the importance of a well-organized, well-articulated strategically directed effort. In the absence of any or all of these preconditions, federal investment will yield less than it could. Therefore, a well-developed and clearly articulated strategic research plan, as called for above, should be a precondition for any new federal appropriations.

Initial federal appropriations on the order of recent spending on desalination research (total appropriations of about $25 million annually, as in fiscal years 2005 and 2006) should be sufficient to make good progress toward the overall research goals if the funding is strategically directed toward the proposed research topics as recommended in this report. Annual federal appropriations of $25 million, properly allocated, should be sufficient to have an impact in the identified priority research areas, given the context of expected state and private-sector funding. This level of federal funding is also consistent with NRC (2004a), which recommended annual appropriations of $700 million for research supporting the nation's entire water resources research agenda. Reallocation of current spending will be necessary to address topics that are currently underfunded. If current research funding is not reallocated, the overall desalination research and development budget will need to be enhanced. Nevertheless, support for the research agenda stated here should not come at the expense of other high-priority water resource research topics, such as those identified in *Confronting the Nation's Water Problems: The Role of Research* (NRC, 2004a).

Environmental research should be emphasized up front in the research agenda. At least 50 percent of the federal funding for desalination research should initially be directed toward environmental research. Environmental research, including Goal 1 and the Goal 1 and 2 crosscuts, should be addressed, because these have the potential for the greatest impact in overcoming current roadblocks for desalination and making desalination an attractive water supply alternative. Research funding in support of Goal 2 should be directed strategically toward research topics that are likely to make improvements against benchmarks set by the best current technologies for desalination. The best available technologies for desalination at the time of this writing are benchmarked in Chapter 4. Research proposals should make the case as to how and to what degree the proposed research can advance the state of the art in desalination. An emphasis should be placed on energy benchmarks because reductions in energy result in overall cost savings and have environmental benefits. The majority of the federal funding directed toward Goal 2 should support projects that are in the public interest *and* would not otherwise be privately funded (see Box 8-1), such as some high-risk and long-term research initiatives (e.g., developing novel desalination processes that sharply reduce the primary energy use). Although private industry does make modest investments in high-risk research, it is frequently reluctant to invest in research in the earliest stage of technology creation, when there is extremely low likelihood of success even though there are large potential benefits.

The effectiveness with which federal funds are spent will also depend on certain critical implementation steps, which are outlined in the following section.

Proposal Announcement and Selection

Based on available funding, the opportunity to announce requests for proposals exists for federal agencies, such as the Bureau of Reclamation or the National Science Foundation, or other research institutions that explicitly target one or more research objectives. The principal funding agency should announce a request for proposals as widely as possible to scientists and engineers in municipal and federal government, academia, and private industry. At present, the desalination community is relatively small, but collectively there is a great deal of expertise across the world. International desalination experts and others from related areas of research should be encouraged and given the opportunity to offer innovative research ideas that have the potential to significantly advance the field. Thus, the request for proposals should extend to federal agencies, national laboratories, other research institutions, utilities, and the private sector. Since innovation cannot be preassigned, broad solicitations for proposals should include a provision for unsolicited investigator-initiated research proposals.

To achieve the objectives of the research agenda, proposals should be selected through a rigorous independent peer-review process (NRC, 2002b) irrespective of the agency issuing the request for proposals. A rotating panel of independent, qualified reviewers should be appointed based on their relevant expertise in the focal areas. The process should allow for the consideration and review of unsolicited proposals, as long as their research goals meet the overall research goals. Proposal funding should be based on the quality of the proposed work, the degree to which the proposed research can advance the state of the art in desalination or otherwise contribute toward the research goals, prior evidence of successful research, and the potential for effective publication or dissemination of the research findings.

CONCLUSIONS AND RECOMMENDATIONS

A strategic national research agenda has been conceived that centers around two overarching strategic goals for further research in desalination: (1) to understand the environmental impacts of desalination and develop approaches to minimize these impacts relative to other water supply alternatives and (2) to develop approaches to lower the financial costs of desalination so that it is an attractive option relative to other alternatives in locations where traditional sources of water are inadequate. A research agenda is proposed in this chapter in support of these two goals (see Box 8-1). Several recommendations for implementing the proposed research agenda follow.

A coordinated strategic plan should be developed to ensure that future federal investments in desalination research are integrated and prioritized and address the two major goals identified in this report. The strategic application of federal funding for desalination research can advance the implementation of desalination technologies in areas where traditional sources of water are inadequate. Responsibility for developing the plan should rest with the OSTP, which should use the recommendations of this report as a basis for plan development. Initial federal appropriations on the order of recent spending on desalination research (total appropriations of about $25 million annually) should be sufficient to make good progress toward these goals, when complemented by ongoing nonfederal and private-sector desalination research, if the funding is directed toward the proposed research topics as recommended in this chapter. Reallocation of current federal spending will be necessary to address currently underfunded topics. If current federal research and development funding is not reallocated, new appropriations will be necessary. However, support for the research agenda stated here should not come at the expense of other high-priority water resource research topics. Five years into the implementation of this plan, the OSTP should evaluate the status of the plan, whether goals have been met, and the need for further funding.

Environmental research should be emphasized up front when implementing the research agenda. Uncertainties regarding environmental impacts and ways to mitigate these impacts are one of the largest hurdles to implementation of desalination in the United States, and research in these areas has the greatest potential for enabling desalination to help meet future water needs in communities facing water shortages. This environmental research includes work to understand environmental impacts of desalination intakes and concentrate management, the development of improved intake methods to minimize impingement and entrainment, and cost-effective concentrate management technologies.

Research funding in support of reducing the costs of desalination (Goal 2) should be directed strategically toward research topics that are likely to make improvements against benchmarks set by the best current technologies for desalination. Because the private sector is already making impressive strides toward Goal 2, federal research funding should emphasize the long-term and high-risk research that may not be attempted by the private sector *and* that is in the public interest, such as research on novel technologies that significantly reduce prime energy use.

Wide dissemination of requests for proposals to meet the goals of the research agenda will benefit the quality of research achieved. Requests for proposals should extend to federal agencies, national laboratories, research institutions, utilities, other countries, and the private sector. Investigator-driven research through unsolicited proposals should be permitted throughout the proposal process. Proposals should be peer-reviewed and based on quality of research proposed, the potential contribution, prior evidence of successful research, and effective dissemination.

References

Abart, E. 1993. Decision-making strategy for handling biocides. Experience at Yuma Desalting Plant. Desalination 97:437-442.

Abdel-Jawad, M. and M. Al-Tabtabaei. 1999. Impact of current power generation and water desalination activities on Kuwaiti marine environment. Paper presented at the IDA World Congress, San Diego, 1999.

Abitoye, J. O., P. Mukherjee, and K. Jones. 2005. Ion Implantation: Effect on Flux and Rejection Properties of NF Membranes. Environmental Science and Technology 39:6487-6493.

Adler, P. R., S. J. del Grosso, and W. J. Parton. 2007. Life-cycle assessment of net greenhouse-gas flux for bioenergy cropping systems. Ecological Applications 17:675-691.

Ahmed, M., A. Arakel, D. Hoey, M. R. Thumarukudy, M. F. A. Goosen, M. Al-Haddabi, and A. Al-Belushi. 2003. Feasibility of salt production from inland RO desalination plant reject brine: A case study. Desalination 158:109-117.

Al-Ahmad, M. and F. A. Aleem. 1993. Scale formation and fouling problems effect on the performance of MSF and RO desalination plants in Saudi Arabia. Desalination 93:287-310.

Al-Shammiri, M., M. Safar, and M. Al-Dawa. 2000. Evaluation of two different antiscalants in real operation at the Doha research plant. Desalination 128:1-16.

Alcamo, J., P. Döll, F. Kaspar, and S. Siebert. 1997. Global change and global scenarios of water use and availability: An application of Water GAP 1.0. Kassel, Germany: University of Gesamthochsch.

Allam, J., G. K. Pearce, and K. Chida. 2003. Ultrafiltration Pre-treatment to RO: Trials at Kindasa Water Services, Jeddah, Saudi Arabia. Paper presented at the 2003 IDA Conference in Bahamas.

Almulla, A. 2002. Developments in high recovery brackish water desalination plants as part of the solution to water quality problems. Desalination 153:237-243.

Alonitis, S. A., K. Kouroumbas, and N. Vlachakis. 2003. Energy consumption and membrane replacement cost for seawater RO desalination plants. Desalination 157:151-158.

Al-Sofi, M. A. K. 2001. Seawater desalination—SWCC experience and vision. Desalination 135:121-139.

Andijani, I., S. Ahmad, and A. Malik. 2000. Corrosion behavior of titanium metal in the presence of inhibited sulfuric acid at 50°C. Desalination 129:45-51.

Anne, C. O., D. Trebouet, P. Jaounen, and F. Quemeneur. 2001. Nanofiltration of seawater: Fractionation of mono- and multi-valent cations. Desalination 140:67-77.

AWWA (American Water Works Association). 1999. Water Quality and Treatment: A Handbook of Community Water Supplies, 5th edition. New York: McGraw-Hill Publishers.

AWWA Membrane Residuals Management Subcommittee. 2004. Committee report: Current perspectives on residuals management for desalting membranes. Journal AWWA 96:73-89.

AWWARF, Lyonnaise des Eaux, Water Research Commission of South Africa. 1996. Water Treatment Membrane Processes. J. Mallevialle, P. E. Odendaal, and M. R., Wiesner, eds. New York: McGraw-Hill Publishers.

Baig, M. and A. A. A. Kutbi. 1998. Design features of a 20mgd SWRO desalination plant, Al Jubail, Saudi Arabia. Desalination 118:5-12.

Banat, F. A., and J. Simandl. 1998. Desalination by membrane distillation: A parametric study. Separation Science and Technology 33(2):201-226.

Baumol, W. J., and W. E. Oates. 1988. The Theory of Environmental Policy, 2nd edition. Cambridge, UK: Cambridge University Press.

Bay, S., and D. Greenstein. 1993. Toxic Effects of Elevated Salinity and Desalination Waste Brine. Southern California Coastal Water Research Projects Annual Report. Available online at: http://www.scwrp.org/pubs/annrpt/92/93. Accessed April 26, 2007.

Beck, R. W. 2002. Alternative Project Delivery: Texas Water Development Board. Available online at: http://www.twdb.state.tx.us/rwpg/rpgm_rpts/2000483347.pdf. Accessed March 28, 2007.

Belfer, S., Y. Purinson, R. Fainshtein, Y. Radchenko, and O. Kedem. 1998. Surface modification of commercial composite polyamide reverse osmosis membranes. Journal of Membrane Science 139:175-181.

Bell, R. 2006. Dana Point Ocean Desalination Project. Presentation given at meeting of the NRC Committeee on Advancing Desalination Technology, Irvine, CA.

Birkett, J. 2003. Desalination Activities in England during the Late 17th Century. International Water History Association: Alexandria, Egypt.

Birkett, J., and R. Truby. 2007. A Figure of Merit for Appreciating Improvements in RO Membrane Performance. International Desalination Association News, Vol. 16.

Blake, N. J., C. W. Dye, M. D. Farrell, and M. A. Hammond. 1996. Effect of disposal of seawater desalination discharges on near shore benthic communities. Phase 1 Report. Southwest Florida Water Management District, Electric Power Research Institute, and University of South Florida.

Blazkowski, C., and G. S. Moreira. 1986. Combined effects of temperature and salinity on the survival and duration of larval stages of *Pagurus criticornis* (Cana) (Crustacea, Paguridae). Journal of Experimental Marine Biology and Ecology 103(11):77-86.

Blumberg, A. F., and G. L. Mellor. 1987. A description of a three-dimensional coastal ocean circulation model. Pp. 1-16 in Three-Dimensional Coastal Ocean Models, N. S. Heaps, ed. Washington, D.C.: American Geophysical Union.

Boland, J. J., B. Dziegielewski, D. D. Bauman, and E. M. Opitz. 1984. Influence of Price and Rate Structures on Minimum and Maximum Water Use. Carbondale, IL: Planning Management Consultants, Ltd.

Bolhuis, P. G., and D. Chandler. 2000. Transition path sampling of cavitation between molecular scale solvophobic surfaces. Journal of Chemical Physics 113:8154-8160.

Bond, R., and V. Veerapaneni. 2007. Zero Liquid Discharge for Inland Desalination. AwwaRF Report No. 3010.

Borgina, M., S. Nielsen, A. Engel, and P. Agre. 1999. Cellular and molecular biology of the aquaporin water channels. Annual Review of Biochemistry 68:425-458.

Bourounia, K., J. C. Deronzierb, and L. Tadrista. 1999. Experimentation and modelling of an innovative geothermal desalination unit. Desalination 125:147-153.

Bousher, A., P. Brimblecombe, and D. Midgley. 1989. Kinetics of reactions in solutions containing monochloramine and bromide. Water Research 23:1049-1058.

Brackish Groundwater Manual for Texas Regional Water Planning Groups. 2003. LBG-Guyton Associates in association with NRS Consulting Engineers, prepared for the Texas Water Development Board, February.

Brian, P. L. T. 1966. Mass transport in reverse osmosis. In Desalination by Reverse Osmosis. Cambridge, MA: MIT Press.

Brining, D. L., W. C. Lester, W. N. Jessee, D. A. O'Leary, S. Bouregois, and M. S. Salam. 1981. Development of environmental standards for combined desalination/power generating stations in the Arabian Gulf region. Desalination 39:255-260.

Brocksen, R. W., and R. E. Cole. 1972. Physiological responses of three species of fishes to various salinities. Journal of the Fisheries Research Board of Canada 29:399-405.

Brown, T. C. 1999. Past and future freshwater use in the United States. Washington, D.C.: U.S. Department of Agriculture, Forest Service.

Brustlin, V. H. 2007. Annual Monitoring Report for the T. Mabry Carlton, Jr. Memorial Reserve Wellfield Water Year 2006 Final Report.

Burashid, K. 1992. Bahrain's Seawater Intakes and Pretreatment Systems. Desalination and Water Reuse 22:44-49.

Buros, O. K. 2000. The ABCs of Desalting, 2nd edition. Topsfield, MA: International Desalination Association.

Buros, O. K., R. B. Cox, I. Nusbaum, A. M. El-Nashar, and R. Bakish. 1980. The USAID Desalination Manual. Englewood Cliffs, NJ: IDEA Publications.

Burroughs, R. 1999. When stakeholders choose: Process, knowledge, and motivation in water quality decision. Society and Natural Resources 12:797-809.

Busch, M., and W. E. Mickols. 2004. Reducing energy consumption in seawater desalination. Desalination 165:299-312.

Cadotte, E. 1977. Reverse Osmosis Membrane. U.S. Patent 4,039,440.

California Coastal Commission. 2004. Seawater Desalination and the California Coastal Act. Available online at: http://www.coastal.ca.gov/energy/14a-3-2004-desalination.pdf. Accessed December 27, 2007.

California Department of Public Health. 2007. Drinking Water Notification Levels and Response Levels: An Overview. Available online at: http://www.cdph.ca.gov/certlic/drinking water/notofocationlevels/notificationlevels.pdf. Accessed January 4, 2008.

California Department of Water Resources. 2005. A Framework for Action: The California Water Plan Update. Bulletin 160-05. Available online at: http://www.waterplan.water.ca.gov/docs/

cwpu2005/cwphighlights/highlights.pdf. Accessed March 14, 2008.

California Energy Commission. 2005a. California's Water Energy Relationship. CEC-700-2005-011-SF. Sacramento, CA.

California Energy Commission. 2005b. Integrated Energy Policy Report 2005. CEC-100-2005-007-CMF. Sacramento, CA.

California Water Desalination Task Force. 2003. Synthesis Paper—Management Practices for Feedwater Intakes and Concentrate Disposal for Seawater Desalination. Available online at: http://www.owue.water.ca.gov/recycle/desal/Docs/Management PracticesIntake_Disposal.doc. Accessed November 16, 2007.

Cao, Y., A. W. Bark, and W. P. Williams. 1996. Measuring the responses of macroinvertebrate communities to water pollution: A comparison of multivariate approaches, biotic and diversity indices. Hydrobiologia 341:1-19.

Carson, R. T., and R. C. Mitchell. 1987. Economic value of reliable water supplies for residential water users in the State Water Project services area. Prepared for the Metropolitan Water District of Southern California. Los Angeles, CA.

Celere, M., and C. Gostoli. 2002. The heat and mass transfer phenomena in osmotic membrane distillation. Desalination 147:133-138.

CEQ (Council on Environmental Quality). 1991. Environmental Quality: 21st Annual Report. Washington, D.C.: CEQ.

Charmentier, G., and M. Charmentier-Daures. 1994. Ontogeny of osmoregulation and salinity tolerance in the isopod crustacean *Sphaeroma serratum*. Marine Ecology Progress Series 114:93-102.

Chesher, R. H. 1975. Biological impact of a large-scale desalination plant at Key West Florida. Pp. 99-181 in Tropical Marine Pollution, E. J. Ferguson Wood and R. E. Johannes, eds. New York: Elsevier Scientific.

Childress, A. E., and M. Elimelech. 2000. Relating nanofiltration membrane performance to membrane charge (electrokinetic) characteristics. Environmental Science and Technology 34:3710-3716.

Civan, F. 2000. Reservoir Formation Damage, Fundamentals, Modeling, Assessment, and Mitigation. Houston, TX: Gulf Publishing Company.

Cohen, R., B. Nelson, and G. Wolff. 2004. Energy Down the Drain: The Hidden Costs of California's Water Supply. New York: National Resources Defense Council and Oakland, CA: Pacific Institute for Studies in Development, Environment, and Security.

Colby, B. G. 1997. Markets as a response to water scarcity: Policy challenges and economic implications. Pp. 211-232 in Advances in the Economics of Environmental Resources, D. Hall, ed. Greenwich, CT: JAI Press.

Cooley, H., P. H. Gleick, and G. Wolff. 2006. Desalination, With a Grain of Salt: A California Perspective. Oakland, CA: Pacific Institute for Studies in Development, Environment, and Security.

Cornes, R., and T. Sandler. 1996. The Theory of Externalities, Public Goods and Club Goods, 2nd edition. Cambridge, UK: Cambridge University Press.

Cotruvo, J. A. 2005. Water desalination processes and associated health and environmental issues. Water Conditioning and Purification, January:13-17.

CPUC (California Public Utilities Commission). 2001. California Standard Practice Manual: Economic Analysis of Demand-Side Programs and Projects. Sacramento, CA: CPUC.

Davis, T. 2006. Zero Discharge Seawater Desalination: Integrating the Production of Freshwater, Salt, Magnesium, and Bromine. USBR Desalination and Water Purification Research and Development Program Report No. 111. Washington, D.C.: U.S. Bureau of Reclamation. Available online at http://www.usbr.gov/pmts/water/publications/reportpdfs/report111.pdf. Accessed March 14, 2008.

De la Cruz, R. V., and C. B. Raleigh. 1972. Absolute stress measurements at the Rangely Anticline, Northwestern Colorado. International Journal of Rock Mechanics and Mining Science & Geometrics Abstracts 9:625-634.

DOI (U.S. Department of the Interior). 2002. Report to Congress on Concepts for National Assessment of Water Availability and Use. U.S. Geological Survey Circular. 1223. Reston, VA: USGS.

Douglas, M., D. Gasper, S. Ney, and M. Thompson. 1998. Human needs and wants. Pp. 195-263 in Human Choices and Climate Change, Vol. 1, S. Rayner and E. Malone, eds. Columbus, OH: Battelle Publishing.

Dow Chemical Company. 1994. Cleaning Procedures for FILMTEC FT 30 Membranes.

Dreizen, Y. 2006. Ashkelon seawater desalination project—off-takers self costs, supplied water costs, total costs and benefits. Desalination 190:104-116.

Drioli, E., E. Curcio, A. Criscuoli, and G. DiProfio. 2004. Integrated system for recovery of $CaCO_3$, NaCl and $MgSO_4$-$7H_2O$ for nanofiltration retentate. Journal of Membrane Science 219:27-38.

Dudek and Associates. 2005. Environmental impact report for precise development plan and desalination plant. EIR 03-05-SCH#2004041081. Prepared for the City of Carlsbad, CA.

Duirk, S. E., and R. L. Valentine. 2007. Bromide oxidation and formation of dihaloacetic acids in chloraminated water. Environmental Science and Technology 41:7047-7053.

DuPont. 1994. PERMASEP Products engineering manual, bulletin 1020, 2010 and 4010. Wilmington, DE: DuPont Company.

Dyksen, J., C. Spencer, R. Hoehn, J. Clement, J. Brandt-Edwards, M. Friedman, A. Hanson, R. Spellecacy, M. McGuire, P. Singer, and A. Camper. 2007. Long-Term Effects of Disinfection Changes on Water Quality. AWWARF Report 91169.

EBMUD (East Bay Municipal Utility District). 2005. Urban Water Management Plan 2005. Oakland, CA: EBMUD.

Ebrahim, S., M. Abdel-Jawad, and M. Safar. 1995. Conventional pretreatment system for the Doha Reverse Osmosis Plant: Technical and economic assessment. Desalination 102:179-187.

Edison Electrical Institute. 2005. Industry Statistics. Available online at: http://www.eei.org/industry_issues/industry_overview_and_statistics/industry_statistics/index.htm#generation. Accessed August 8, 2007.

Einav, R., K. Harussi, and D. Perry. 2002. The footprint of desalination processes on the environment. Desalination 152:141-154.

El-Bourawi, M. S., Z. Ding, M. Khavet, and R. Ma. 2006. A framework for better understanding membrane distillation separation process. Journal of Membrane Sciences 285:4-29.

EPA (U.S. Environmental Protection Agency). 1979. National secondary drinking water regulations, final rule. Federal Register 44 (140): 42195-42202.

EPA. 1985. Abient Water Quality Criteria for Copper—1984. Office of Water Regulations and Standards. Washington, D.C.: EPA 440/5-84-031.

EPA. 1986. Abient Water Quality Criteria for Nickel—1986. Office of Water Regulations and Standards. Washington, D.C.: EPA 440/5-86-004.

EPA. 1991. "CORMIX2: An Expert System for Hydrodynamic Mixing Zone Analysis of Conventional and Toxic Submerged Multiport Diffuser Discharges", Technical Report EPA/600/3-91/073, U.S. Environmental Research Laboratory, Athens, Georgia.

EPA. 2005. A Regulators' Guide to the Management of Radioactive Residuals from Drinking Water Treatment Technologies. EPA 816-R-05-004. Available online at: http://www.epa.gov/radiation/docs/tenorm/816-r-05-004.pdf. Accessed December 21, 2007.

EPA. 2006. Health Effects Support Document for Boron. Office of Water. Washington, D.C.: EPA-822-R-06-005.

EPA. 2007a. Regulatory Support Documents for Selected Contaminants from the Second Drinking Water Contaminant Candidate List

(CCL2). EPA Report 815-D-06-007. Washington, D.C.: EPA. Available online at: http://epa.gov/safewater/ccl/pdfs/reg_determine2/report_ccl2-reg2_supportdocument_ch03_boron.pdf. Accessed February 21, 2008.

EPA. 2007b. Underground Injection Control Program. Available online at: http://www.epa.gov/safewater/uic/index.html. Accessed April 8, 2008.

EPA. 2007c. Drinking Water State Revolving Fund (DWSRF) Allotments. Washington, D.C.: EPA. Available online at: http://.epa.gov/safewater/dwsrf/alloatments/bascic information.html. Accessed February 22, 2008.

Evans, D. M. 1966. The Denver area earthquakes and the Rocky Mountain Arsenal disposal well. The Mountain Geologist 3(1):23-26.

European Commission. 1998. Desalination Guide Using Renewable Energies. Brussels: European Commission.

European Commission. 2007. Global Change and Ecosystems: EU Research for the Environment, Catalogue of FP6 Projects. Available online at: http://ec.europa.eu/research/environment/pdf/global_change_ecosystem.pdf. Accessed December 11, 2007.

Farmer, J. C., J. Richardson, D. Fix, S. Thomson, and S. May. 1996. Desalination with Carbon Aerogel Electrodes. Lawrence Livermore National Laboratory (LLNL) Report No. URCL-ID-125298 Rev 1.

Ferraris, J. D., K. Fauchald, and B. Kensley. 1994. Physiological responses to fluctuation in temperature of salinity in invertebrates. Adaptations of *Alpheus viridari* (Decapoda, Crustacea), *Terebellides parva* (Polychaeta) and *Golfinigia cylindrata* (Sipunculida) to the mangrove habitat. Marine Biology 120:397-406.

Feth, J. H. 1965. Preliminary map of the conterminous United States showing depth to and quality of shallowest ground water containing more than 1,000 part per million dissolved solids. U.S. Geological Survey Hydrologic Investigation Atlas HA-199. Reston, VA: USGS.

FilmTec. 2000. FILMTEC membranes tech manual excerpts. FilmTec Corporation, The Dow Chemical Company. Available online at: http://www.dow.com/liquidseps.

FitchRatings. 2007. Revenue Criteria Report: Water and Sewer Revenue Bond Rating Guidelines. New York: Fitch Inc. Available online at: http://www.fitchratings.com/corporate/login/setSessionVars.cfm?userIdParam=mstoever&SCRIPT_NAME=/corporate/reports/report_frame.cfm&QUERY_STRING=rpt_id=322724&sector_flag=&marketsector=3&detail=. Accessed November 16, 2007.

Forster, S .J. 1998. Osmotic stress tolerance and osmoregulation of intertidal and subtidal nematodes. Journal of Experimental Marine Biological Ecology 224:109-125.

Freeze, R. A., and J. A. Cherry. 1979. Groundwater. Upper Saddle River, NJ: Prentice-Hall.

Frenkel, V. 2004. Desalination Methods, Technology and Economics. Presentation at 2004 Desalination Conference, Santa Barbara, CA. Available online at: http://www.idswater.com/Common/Paper/Paper_90/Val%20Frenkle%20papaer.pdf.

Friedman, E., and H. Moulin. 1999. Three methods to share joint costs or surplus. Journal of Economic Theory 87:275-331.

Gacia, E., O. Invers, M. Manzanera, E. Ballesteros, and J. Romero. 2007. Impact of the brine from a desalination plant on a shallow seagrass (*Posidonia oceanica*) meadow. Estuarine, Coastal and Shelf Science 72:579-590.

Galloway, D. L., D. R. Jones, and S. E. Ingebritsen. 1999. Land subsidence in the United States. U.S. Geological Survey Circular 1182. Reston, VA: USGS. Available online at: http://pubs.usgs.gov/circ/circ1182/. Accessed December 11, 2007.

Galloway, M., and J. Mahoney. 2004. Ultrafiltration for seawater reverse osmosis pretreatment. Membrane Technology 2004:5-8.

García-Rodríguez, L. 2003. Renewable energy applications in desalination: State of the art. Solar Art 75:381-393.

Gaston, G. R., and J. C. Young. 1992. Effects of contaminants on macrobenthic communities in the upper Calcasieu Estuary, Louisiana. Bulletin of Environmental Contaminants and Toxicology 37:922-928.

Geisler, P., W. Krumm, and T. A. Peters. 2001. Reduction of the energy demand for seawater RO with the pressure exchange system PES. Desalination 135:205-210.

German Aerospace Center (Deutsches Zentrum für Luft—und Raumfahrt or DLR). 2007. Concentrating Solar Power for Seawater Desalination. Prepared for the German Federal Ministry for the Environment, Nature Conservation and Nuclear Safety. Available online at: http://www.dll.de/tt/aqua-csp. Accessed on April 8, 2008.

Gilron, J., S. Belfer, P. Väisänen, and M. Nysrtöm. 2001. Effects of surface modification on antifouling and performance properties of reverse osmosis membranes. Desalination 140: 167-179.

Glater, J., M. R. Zachariah, S. B. McCray, and J. W. McCutchan. 1983. Reverse osmosis membrane sensitivity to ozone and halogen disinfectants. Desalination 48:1-16.

Gleick, P. H. 1997. Water 2050: Moving Toward A Sustainable Vision for the Earth's Fresh Water. Oakland, CA: Pacific Institute for Studies in Development, Environment, and Security.

Gleick, P. H. 2002. Soft water paths. Nature 418:373.

Gleick, P. H. 2003a. Global freshwater resources: Soft-path solutions for the 21st century. Science 302:1524-528.

Gleick, P. H. 2003b. Water use. Annual Review of Environmental Resources 28:275-314.

Gleick, P. H., G. Wolff, E. L. Chalecki, and R. Reyes. 2002. The New Economy of Water: The Risks and Benefits of Globalization and Privatization of Fresh Water. Oakland, CA: Pacific Institute for Studies in Development, Environment, and Security.

Gleick, P. H., D. Haasz, C. Henges-Jeck, V. Srinivasan, G. Wolff, K. K. Cushing, and A. Mann. 2003. Waste Not, Want Not: The Potential for Urban Water Conservation in California. Oakland, CA: Pacific Institute for Studies in Development, Environment, and Security.

Grattan, S., C. Grieve, J. Poss, P. Robinson, D. Suarze, and S. Benes. 2004. Evaluation of salt-tolerant forages for sequential water re-use systems. Implications for ruminant mineral nutrition. Agriculture Water Management 70:137-150.

Grieve, C., J. Poss, S. Grattan, D. Suarze, S. Benes, and P. Robinson. 2004. Evaluation of salt-tolerant forages for sequential reuse systems. Agriculture Water Management 70:121-135.

Griffin, R. C., and J. W. Mjelde. 2000. Valuing water supply reliability. American Journal of Agricultural Economics 82:414-426.

GWI (Global Water Intelligence). 2006a. Desalination Markets 2007: A Global Forecast. Oxford, UK: Media Analyics Ltd.

GWI. 2006b. 19th IDA Worldwide Desalting Plant Inventory. Oxford, UK: Media Analytics Ltd.

GWI. 2007. Desalination Tracker 8:37-48. Oxford, UK: Media Analytics Ltd.

Hamed, O. A. 2005. Overview of hybrid desalination systems—current status and future prospects. Desalination 186:207-214.

Hamed, O. A., M. Al-Sofi, G. Mustafa, and A. Dalvi. 1999. The performance of different and anti-scalants in multi-stage flash distillers. Desalination 123:185-194.

Hamed, O. A., M. Al-Sofi, M. Imam, K. Mardouf, A. Al-Mobayed, and A. Ehsan. 2000. Evaluation of polyphosphonate antiscalant at a low does rate in the Al-Jubail Phase II MSF plant, Saudi Arabia. Desalination 128: 275-280.

Hamed, O. A., M. A. K. Al-Sofi, G. M. Mustafa, K. Ba-Mardouf, and H. Al-Washmi. 2002. Power/Water Cogeneration Cycles Analysis -

Part 1, Design Data Theoretical Analysis. Paper presented at IDA World Congress, Saudi Arabia, 2002.

Hamieh, B. M., J. R. Beckman, and M. D. Ybarra. 2001. Brackish and seawater desalination using a 20 foot dewvaporation tower. Desalination 140(3):217-226.

Hammond, M. A., N. J. Blake, C. W. Dye, P. Hallock-Muller, M. E. Luther, D. A. Tomasko, and G. Vargo. 1998. Effect of Disposal of Seawater Desalination discharges on near shore benthic communities, Phase Two Report. Southwest Florida Water Management District, Electric Power Research Institute, University of South Florida.

Hanafi, A. 2002. Cogeneration of Power and Desalinated Water. Paper presented at IDA World Congress, Egypt, 2002.

Haneman, W. M. 2006. The economic conception of water. In Water Crisis: Myth or Reality, P. P. Rogers, M. R. Llamas, and L. Martinez Cortina, eds. London: Taylor and Francis.

Hannam, K. M., L. W. Oring, and M. P. Herzog. 2003. Impacts of salinity on growth and behavior of American avocet chicks. Waterbirds 26:119-125.

Harbaugh, A. W. 2005. MODFLOW-2005, the U.S. Geological Survey modular ground-water model—the ground-water flow process. U.S. Geological Survey Techniques and Methods 6-A16. Reston, VA: USGS.

Heberer, T., B. Fuhrmann, K. Schmidt-Baumler, D. Tsipi, V. Koutsouba, and A. Hiskia. 2001. Occurrence of pharmaceutical residues in sewage, river, ground and drinking water in Greece and Germany. In Pharmaceuticals and Personal Care Products in the Environment: Scientific and Regulatory Issues, Symposium Series 791, D. B. Daughton and T. Jones-Lepp, eds. Washington, D.C.: American Chemical Society.

Hem, J. D. 1986. Study and interpretation of the chemical characteristics of natural water. U.S. Geological Survey Water-Supply Paper 2254. Reston, VA: USGS.

Hinds, B. J., N. Chopra, T. Rantell, R. Andrews, V. Gavalas, and L. G. Bachas. 2004. Aligned multiwalled carbon nanotube membranes. Science 303:62-65.

Hinkebein, T. E., and M. K. Price. 2005. Progress with the desalination and water purification technologies U.S. roadmap. Desalination 182:19-28.

Hoffman, D., H. M. Ohlendorf, and T. W. Aldrich. 1988. Selenium teratogenesis in natural populations of aquatic birds in Central California. Archives of Environmental Contamination and Toxicology 17:519-525.

Holt, J. K., H. G. Park, Y. Wang, M. Stadermann, A. B. Artyukhin, C. P. Grigoropoulos, A. Noy, and O. Bakajin. 2006. Fast mass transport through sub-2-nanometer carbon nanotubes. Science 312:1034-1037.

Höpner, T. 1999. A procedure for environmental impact assessments (EIA) for seawater desalination plants. Desalination 124:1-12.

Höpner, T., and S. Lattemann. 2002. Chemical impacts from seawater desalination plants—a case study of the Red Sea. Desalination 152:133-140.

Höpner, T., and J. Windelberg. 1996. Elements of environmental impact studies on coastal desalination plants. Desalination 108:11-18.

Horne, M. T., N. J. Finley, and M. D. Sprenger. 1999. Polychlorinated biphenyl- and mercury-associated alterations on benthic invertebrate community structure in a contaminated salt marsh in Southeast Georgia. Archives of Environmental Contamination and Toxicology 37:317-325.

Hsieh, P. A., and J. D. Bredehoeft. 1981. A reservoir analysis of the Denver earthquakes: A case of induced seismicity. Geophysical Research Letters 86:903-920.

Huff, G. F. 2004a. An overview of the hydrogeology of saline ground water in New Mexico. Las Cruces, NM: New Mexico Water Resources Research Institute.

Huff, G. F. 2004b. Review of knowledge on the occurrence, chemical composition, and potential use for desalination of saline ground water in Arizona, New Mexico, and Texas with a discussion of potential future study needs. U.S. Geological Survey Open-File Report 2004-1197. Reston, VA: USGS.

Hummer, G., J. C. Rasalah, and J. P. Noworyta. 2001. Water conduction through the hydrophobic channel of a carbon nanotube. Nature 414:188-190.

Hussain, A. and A. Ahmed. 1998. The Addur SWRO desalination plant towards a full plant production. Desalination 120:143-152.

Hutson, S., S. Nancy, L. Barber, J. F. Kenny, K. S. Linsey, D. S. Lumia, and M. A. Maupin. 2004. Estimated Use of Water in the United States in 2000. U.S. Geological Survey, USGS Circular 1268. Reston, VA: USGS. Available online at: http://pubs.usgs.gov/circ/2004/circ1268/. Accessed March 14, 2008.

Hyung, H., and J.-H. Kim. 2006. A mechanistic study on boron rejection by sea water reverse osmosis membranes. Journal of Membrane Science 286:269-278.

IAEA (International Atomic Energy Agency). 2005. Optimization of the coupling of nuclear reactors and desalination systems. IAEA-TECDOC-1444. Vienna, Austria: IAEA.

Imam, M., R. Al-Rasheed, I. Al-Tisan, A. Abdulgader, G. Mustafa, A. S. Al-Mobayed, A. Ehsan, and D. Brose. 2000. Studies on a polyoxypropylene glycol-based antifoaming agent in MSF plants. Desalination 129:187-196.

Institute of Medicine (IOM). 2000. Dietary Reference Intakes for Vitamin A, Vitamin K, Arsenic, Boron, Chromium, Copper, Iodine, Iron, Manganese, Molybdenum, Nickel, Silicon, Vanadium, and Zinc. Washington, D.C.: National Academies Press.

Iso, S., S. Suizu, and A. Maejima. 1994. The lethal effect of hypertonic solutions and avoidance of marine organisms in relation to discharged brine from a destination plant. Desalination 97:389-399.

Jacangelo, J., and J. Grounds. 2004. Properly designed conventional pretreatment can provide suitable water quality for RO. International Desalination & Water Reuse Quarterly 13:16-23.

Jenkins, S. A., and J. B. Graham. 2006. Oceanographic considerations for desalination plants in Southern California Coastal Waters, Parts 1 and 2. Presentation given at meeting of NRC Committee on Advancing Desalination Technology, Irvine, CA.

Jeong, B.-H., E. M. V. Hoek, Y. Yan, A. Subramani, X. Huang, G. Hurwitz, A. K. Ghosh, and A. Jawor. 2007. Interfacial polymerization of thin film nanocomposites: A new concept for reverse osmosis membranes. Journal of Membrane Science 294:1-7.

Johnson, R. L., P. G. Tratnyek, R. Miehr, R. B. Thoms, and J. Z. Bandstra. 2005. Reduction of hydraulic conductivity and reactivity in zero-valent iron columns by oxygen and TNT. Ground Water Monitoring & Remediation 25:129-136.

Johnston, L., and S. H. Williamson. 2002. The Annual Real and Nominal GDP for the United States, 1789-Present. Oxford, OH: Economic History Services.

Jordahl, J. 2006. Beneficial and Non-traditional Uses of Concentrate. Alexandria, VA: WateReuse Foundation.

Jun, L., B. McClain, A. Seah, and D. Mukhopadhyay. 2004. High-purity watersystem upgrade in Singapore using high efficiency RO. Ultrapure Water 21:27-30.

Kalra, A., S. Garde, and G. Hummer. 2003. Osmotic water transport through carbon nanotube membranes. Proceedings of the National Academy of Sciences 100:10175-10180.

Karajeh, F. 2006. Proposition 50 Chapter 6(a) Water Desalination Grants—2006 Funding Cycle. Sacramento, CA: California Department of Water Resources. Available online at:

http://www.owue.water.ca.gov/recycle/DesalPSP/FinalFundingAwards2006.pdf. Accessed November 20, 2007.

Karry, H. 2006. Prototype Testing Facility for Two Pass Membrane Seawater Desalination Process, Membrane Treatment for Drinking Water and Re-Use Applications: A Compendium of Peer Reviewed Papers. Denver, CO: AWWA.

Karytsasa, C., D. Mendrinosa, and G. Radogloub. 2004. The current geothermal exporation and development of the geothermal field of Milos island in Greece. Geo-Heat Center Quarterly Bulletin 25:17-21.

Kenna, E. N., and A. K. Zander. 2000. Current management of membrane plant concentrate. Denver, CO: AWWARF.

Kenna, E. N., and A. K. Zander. 2001. Survey of membrane concentrate reuse and disposal. Denver, CO: AWWARF.

Khordaugh, H. 1992. Conceptual approach to selection of a control measure for residual chlorine discharge in Kuwait Bay. Environmental Management 3:309-316.

Kiernan, J., and A. von Guttberg. 2005. Half Selection of EDR Desalting Technology Rather than MF/RO for the City of San Diego Water Reclamation Project. Niskuyuna, NY: GE Water & Process Technologies.

KISR (Kuwait Institute for Scientific Research). 2007. Water Resources Division: Mission and Objectives. Available online at: http://www.kisr.edu.kw/divisions_water_mission.asp. Accessed December 11, 2007.

Kollar, K. L., and P. MacAuley. 1980. Water requirements for industrial development. Journal of the American Water Works Association 72:2-9.

Kolpin, D. W., D. A. Goolsby, and E. M. Thurman. 1995. Pesticides in near-surface aquifers—an assessment using highly sensitive analytical methods and tritium. Journal of Environmental Quality 24:1125-1132.

Koomey, J., C. Atkinson, A. Meier, J. E. McMahon, S. Boghosian, C. Dunham, and J. D. Lutz. 1995. The effect of efficiency standards on water use and water heating energy use in the U.S.: A detailed end-use treatment. Energy—The International Journal 20:627.

Lahav, O., and L. Birnac. 2007. Quality criteria for desalinated water following post-treatment. Desalination 207:286-303.

Laine, J. M., J. G. Jacangelo, K. E. Carns, E. W. Cummings, and J. Mallevialle. 1993. Influence of bromide on low pressure membrane filtration for controlling disinfection by-products in surface waters. Journal AWWA 81:61.

Langevin, C. D., W. B. Shoemaker, and W. Guo. 2003. SEAWAT-2000: A Version of MODFLOW-2000 with the Variable-Density Flow

Process and the Integrated MT3DMS Transport Process. U.S. Geological Survey Open-File Report 03-426. Reston, VA: USGS. Available online at: http://water.usgs.gov/ogw/seawat/seawat2k.html. Accessed December 12, 2007.

Lasker, R., R. Tanaza, and L. Chamberlain. 1972. The response of Salton Sea fish eggs and larvae to salinity stress. California Department of Fish and Game 58:58-66.

Latorre, M. 2001. Surface open intake SWRO pilot plant with MF pretreatment. Presentation at 2001 IDA World Congress, Bahrain.

Latorre, M. 2005. Environmental impact of brine disposal on Posidonia seagrasses. Desalination 182:517-524.

Lattemann, S., and T. Höpner. 2003. Seawater Desalination: Impacts of Brine and Chemical Discharge on the Marine Environment. L'Aquila, Italy: Balaban Desalination Publications.

Lattemann, S., and T. Höpner. 2007. Impacts of seawater desalination plants on the marine environment of the Gulf. In Protecting the Gulf's Marine Ecosystems from Pollution, A. Abuzinada, H.-J. Barth, F. Krupp, B. Boer, and T. Abdelsalaam, eds. Switzerland: Birkhauser Verlag.

Leake, S. A. 2007. Land Subsidence from Ground-Water Pumping. Tuscon, AZ: USGS. Available online at: http://geochange.er.usgs.gov/sw/changes/anthropogenic/subside/. Accessed April 26, 2007.

Lieberman, B. 2003. Importance of Energy Recovery Devices in Reverse Osmosis Desalination. Austin, TX: Texas Water Development Board.

Liu, C. C. K., J. W. Park, J. Migita, and G. Qing. 2002. Experiments of a prototype wind-driven reverse osmosis desalination system with feedback control. Desalination 150:277-287.

Loeb, S., and S. Sourirajan. 1963. Seawater Demineralization by means of an osmotic membrane. Advances in Chemistry Series 148:117.

Lonsdale, K., U. Merten, and R. L. Riley. 1965. Transport properties of cellulose acetate osmotic membranes. Journal of Applied Polymer Science 9:1341-1362.

Lu, H., J. C. Walton, and A. H. P. Swift. 2001. Desalination coupled with salinity-gradient solar ponds. Desalination 136:13-23.

Ludwig, H. 2004. Hybrid systems in seawater desalination—practical design aspects, present status, and development perspectives. Desalination 164:1-18.

Luther, M. E., B. Galperin, M. Vincent, D. Burwell, R. Pribble, and A. Janicki. 1998. Potential Effects of Tampa Bay Water Surface Water Projects on Salinity and Circulation in Tampa

Bay: Results of the USF Three-Dimensional Hydrodynamic Model. Prepared for Tampa Bay Water. Tampa, FL: PBS&J.
MacDonnell, L. 1990. Transferring Water Uses in the West. Oklahoma Law Review. 43:119
MacKichan, K. A. 1951. Estimated Water Use in the United States, 1950: U.S. Geological Survey Circular 115. Reston, VA: USGS.
MacKichan, K. A. 1957. Estimated Water Use in the United States, 1955: U.S. Geological Survey Circular 398. Reston, VA: USGS.
MacKichan, K. A., and J. C. Kammerer. 1961. Estimated Use of Water in the United States, 1960: U.S. Geological Survey Circular 456. Reston, VA: USGS.
Magara, Y., A. Tabata, M. Kohki, M. Kawasaki, and M. Hirose. 1998. Development of boron reduction system for sea water desalination. Desalination 118:25-33.
Mahajan, R., D. Q. Vu, and W. J. Koros. 2002. Mixed matrix membrane materials: An answer to the challenges faced by membrane based gas separations today? Journal of the Chinese Institute of Chemical Engineers 33:77-86.
Mallevialle, J., P. Odendaal, and M. R. Wiesner, eds. 1996. Membrane Processes in Water Treatment. New York: McGraw-Hill Publishers.
Malmrose, P., J. Lozier, M. Mickley, R. Reiss, J. Russell, J. Schaefer, S. Sethi, J. Manuszak, R. Bergman, and K. Atasi. 2004. Committee Report: Current Perspectives on Residuals Management for Desalting Membranes. Journal AWWA 96:73-87.
McCusker, A. J., B. G. Dove, H. B. Dreyer, S. J. Fletcher, M. A. Hamlin, and K. M. Nordfors. 2007. Intake Arrangement to Mitigate the Environmental Effects of Surface Water Withdrawal. Presentation at IDA World Congress, Gran Canaria, 2007.
McCutcheon, J. R., R. L. McGinnis, and M. Elimelech. 2005. A novel ammonia-carbon dioxide forward (direct) osmosis desalination process. Desalination 174:1-11.
McCutcheon, J. R., R. L. McGinnis, and M. Elimelech. 2006. Desalination by ammonia-carbon dioxide forward osmosis: Influence of draw and feed solution concentrations on process performance. Journal of Membrane Science 278:114-123.
McGinnis, R. L., and M. Elimelech. 2007. Energy requirements of ammonia carbon dioxide forward osmosis desalination. Desalination 207:370-382.
McHarg, J., and R. Truby. 2004. West coast researchers seek to demonstrate SWRO affordability. Desalination and Water Reuse Quarterly 14:10-18.
MEDRC (Middle East Desalination Research Center). 2007. R&D Projects Request for Proposals. Cyprus, Greece: Research

Promotion Foundation. Available online at: http://www.medrc.org. Accessed December 11, 2007.

Merkel, T. C., B. D. Freeman, R. J. Spontak, Z. He, I. Pinnau, P. Meakin, and A. J. Hill. 2002. Ultrapermeable, reverse-selective nanocomposite membranes. Science 296:519-522.

Mickley, M. C. 2006. Membrane Concentrate Disposal: Practices and Regulation (Second Edition) Report 123. Denver, CO: U.S. Bureau of Reclamation.

Mickley, M. C. 2007. Survey of High Recovery and Zero Liquid Discharge Technologies for Water Utilities. Alexandria, VA: WateReuse Foundation.

Mickley, M. C., R. Hamilton, L. Gallegos, and J. Truesdall. 1993. Membrane Concentrate Disposal. Mickley and Associates, Boulder, CO, AWWARF, and AWWA.

Mielke, J. E. 1999. Desalination R&D: The New Federal Program. Congressional Research Service Report RS20069. Washington, D.C.: Congressional Research Service.

Miller, J. E. 2003. Review of Water Resources and Desalination Technologies. Sandia National Laboratories Report, SAND 2003-0800. Albuquerque, NM: Sandia National Laboratories.

Miller, J. E., and T. M. Mayer. 2005. A Critical Review of Alternative Desalination Technologies. Prepared for the Joint Water Reuse & Desalination Task Force.

Minato, A., and M. Hirai. 2003. Present and future activities of nuclear desalination in Japan. International Journal of Nuclear Desalination 1:259-270.

Mirant Lovett LLC. 2006. Gunderboom MLES Evaluation Study at the Lovett Generating Station, Tomkins, New York, Results of 2005 biological monitoring. Prepared by ASA Analysis & Communication, Inc., Washingtonville, NY, January.

Missimer, T. M. 1994. Water supply development for membrane water treatment facilities. Boca Raton, FL: CRC Press

Morton, A., I. Callister, and N. Wade. 1996. Environmental impacts of seawater distillation and reverse osmosis processes. Desalination 108:1-10.

Mukhopadhyay, A. 1999. Method and apparatus for high-efficiency reverse osmosis operation. U.S. Patent 5,925,255.

Mulder, M. 2004. Basic Principles of Membrane Technology, 2nd edition. New York: Springer-Verlag New York.

National Association of Manufacturers. 1950. Water in industry. Report of the Economic Policy Division Series 36, Appendix C. Washington, D.C.: National Association of Manufacturers.

New Mexico Office of the State Engineer, New Mexico Water Resources Research Institute, U.S. Bureau of Reclamation. 2004. New

Mexico Brackish Groundwater Assessment Program Workshop.. Report of Findings and Evaluations. Presented at meeting held January 14, 2004, Albuquerque, NM. Available online at: http://wrri.nmsu.edu/conf/brackishworkshop/report/NMBWWreport_FINAL.pdf. Accessed November 16, 2007.

NRC (National Research Council). 1989. Irrigation-Induced Water Quality Problems. Washington, D.C.: National Academy Press.

NRC. 1991. Mitigating Losses from Land Subsidence in the United States. Washington, D.C.: National Academy Press.

NRC. 1992. Water Transfers in the West: Efficiency, Equity and the Environment Washington, D.C.: National Academy Press.

NRC. 1993. Soil and Water Quality: An Agenda for Agriculture. Washington, D.C.: National Academy Press.

NRC. 1996. Toxicological and Performance Aspects of Oxygenated Motor Vehicle Fuels. Washington, D.C.: National Academy Press.

NRC. 1997. Safe Water from Every Tap: Improving Water Service to Small Communities. Washington, D.C.: National Academy Press.

NRC. 1999. Hormonally Active Agents in the Environment. Washington, D.C.: National Academy Press.

NRC. 2002a. Estimating Water Use in the United States: A New Paradigm for the National Water Use Information Program. Washington, D.C.: National Academies Press.

NRC. 2002b. The Missouri Ecosystem: Exploring the Prospects for Recovery. Washington, D.C.: National Academies Press.

NRC. 2002c. Privatization of Water Services in the United States. Washington, D.C.: National Academies Press.

NRC. 2004a. Confronting the Nation's Water Problems: The Role of Research. Washington, D.C.: National Academies Press.

NRC. 2004b. Review of the Desalination and Water Purification Technology Roadmap. Washington, D.C.: National Academies Press.

NRC. 2004c. Valuing Ecosystem Services: Toward Better Environmental Decision Making. Washington, D.C.: National Academies Press.

NRC. 2005. Health Implications of Perchlorate Ingestion. Washington, D.C.: National Academies Press.

NRC. 2007. Colorado River Basin Water Management: Evaluating and Adjusting to Hydroclimatic Variability. Washington, D.C.: National Academies Press.

Oakland, W. H. 1987. Theory of public goods. Pp. 485-535 in Handbook of Public Economics, A. Auesrbach and M. Feldstein, eds. New York: Elsevier.

OMB (U.S. Office of Management and Budget). 2003. Circular A-11: Preparation, Submission, and Execution of the Budget. Washington, D.C.: Executive Office of the President. Available online at: http://www.whitehouse.gov/omb/circulars/a11/02toc.html. Accessed February 26, 2008.

O'Neill, K. M., M. H. Schreider, T. M. Glasby, and A. R. Redden. 2007. Lack of epifaunal response to the application of salt for managing the noxious green alga *Caulerpa taxifolia* in a coastal lake. Hydrobiologia 580:135-142.

Pankratz, T. 2005. Advances in desalination technology. International Journal of Nuclear Desalination 2005 1(4):450-455.

Pankratz, T. 2008. ZLD systems are operating challenge. Water Desalination Report 44(7):1.

Pankratz, T., and J. Tonner. 2003. Desalination.com. Houston, TX: Lone Oak Publishing.

Parkhurst, D. L. 1995. Users guide to PHREEQC: A computer program for speciation, reaction-path, advective transport, and inverse geochemical calculations. U.S. Geological Survey Water Resources Investigations Report 95-4227:143. Reston, VA: USGS.

Pearce, G. K. 2007. The case for UF/MF pretreatment to RO in seawater applications. Desalination 203:286-295.

Peng, P., A. G. Fane, and X. Li. 2005. Desalination by membrane distillation adopting a hydrophyllic membrane. Desalination 173(1):45-54.

Perez Talavera, J. L., and J. Quesada Ruiz. 2001. Identification of the mixing processes in brine discharges carried out in Barranco del Toro Beach, south of Gran Canaria (Canary Islands). Desalination 139:277-286.

Peters, T., D. Pinto, and E. Pinto. 2007. Improved seawater intake and pre-treatment system based on Neodren technology. Desalination 203:134-140.

Pilar Ruso, Y. D., J. A. Ossa Carretero, F. Giminez Casalduero, and J. L. Sanchez Lizaso. 2007. Spatial and temporal changes in infaunal communities inhabiting soft-bottoms affected by brine discharges. Marine Environmental Research 64:492-503.

Pillard, D. A., D. L. DuFresna, J. E. Tietge, and J. M. Evans. 1999. Response of mysid shrimp *(Mysidopsis bahia*), sheepshead minnow (*Cyprinodon variegates*) and inland silversides minnow (*Menidia beryllina*) to changes in artificial seawater salinity. Environmental Toxicology and Chemistry 18:430-435.

Plummer, L. N., D. L. Parkhurst, G. W. Fleming, and S. A. Dunkle. 1988. A computer program incorporating Pitzer's equations for calculation of geochemical reactions in brines. U.S. Geological

Survey Water-Resources Investigations Report 88-4153. Reston, VA: USGS.

Posco Steel. 2007. Environmental Performance: Air and Water. Pohang, Korea: Posco Steel. Available online at: http://www.posco.co.kr/homepage/docs/en/sustain/environ/s91d1010050c.html. Accessed December 4, 2007.

Prato, T. E. Schoepke, L. Etchison, T. O'Brien, B. Hernon, K. Perry, and M. Peterson. 2002. Production of high purity water from seawater. Water Quality Products 7:3.

Rachid, M., and C. Abdelwahab. 2005. Ecotoxicological marine impacts from seawater desalination plants. Desalination 182:403-410.

Raftelis, G. A. 1989. The Arthur Young Guide to Water and Wastewater Finance and Pricing. Chelsea, MI: Lewis Publishers.

Rainey, C. J., L. A. Nyquist, R. E. Christensen, P. L. Strong, B. D. Culver, and J. R. Coughlin. 1999. Daily boron intake from the American diet. Journal of the American Diet Association 99:335–340.

Raucher, R., D. Chapman, J. Henderson, M. Hagenstad, J. Rice, J. Goldstein, A. Huber-Lee, B. Hurd, R. Linsky, E. Means, and M. Renwick. 2005. The Value of Water: Concepts, Empirical Evidence, and their Application to Water Management Decisions. Denver, CO: American Water Works Association Research Foundation.

Raventos, N., E. Macpherson, and A. Garcia-Rubics. 2006. Effect of brine discharge from a desalination plant on macrobenthic communities in the NW Mediterranean. Marine Environmental Research 62:1-14.

Reahl, E. R. 2006. Half Century of Desalination with Electrodialysis. GE Water & Process Technologies, General Electric Company.

Redondo, J. and I. Lomax. 1997. Experiences with the pretreatment of raw water with high fouling potential for reverse osmosis plant using FILMTEC membranes. Desalination 110:167-182.

Reiss Environmental, Inc. 2003. Disinfection Residual Stability Study. Internal report for Tampa Bay Water.

Renwick, M. E., and R. D. Green. 2000. Do residential water demand side management policies measure up: An analysis of eight California water agencies. Journal of Environmental Economics and Management 40:37-55.

Reynolds, W. W., D. A. Thomson, and M. E. Casterlin. 1976. Temperature and salinity tolerance of larval California grunion, *Leuristhes tenuis* (Ayres): A comparison with gulf grunion, *L. sardinia* (Jenkins and Evermann). Journal of Experimental Marine Biology and Ecology 24:73-82.

Ricci, J. and E. Bailey. 2007. Smelt Ruling May Cut Into Water Supply. Printed in the Los Angeles Times, September 1, 2007.

Richards, B. S., and A. I. Schäfer. 2003. Photovoltaic-powered desalination system for remote Australian communities. Renewable Energy 28:2013-2022.

Richardson, S. D., T. V. Caughran, P. H. Chen, T. W. Collette, T. L. Floyd, and A. D. Thruston. 1999. Identification of new drinking water disinfection byproducts formed in the presence of bromide. Environmental Science and Technology 33:3378-3383.

Richardson, S. D., M. J. Plewa, E. D. Wagner, R. Schoeny, and D. M. DeMarini. 2007. Occurrence, genotoxicity and carcinogenicity of regulated and emerging disinfection by-products in drinking water: a review and roadmap for research. Mutation Research-Reviews in Mutation Research 636:178-242.

Riley, R. L., R. L. Fox, C. R. Lyons, C. E. Milstead, M. W. Seroy, and M. Tagami. 1976. Spiral-wound polyether amide thin film composite membrane systems. Desalination 19:1113.

Rinne, W. W. 1971. Waste brine disposal from coastal desalination plants. Pp. 325-328 in IEEE Engineering in the Ocean Environment Conference.

Roberts, N. 2004. Public deliberation in an age of direct citizen participation. American Review of Public Administration 34:315-353.

Robinson, M. K. 2007. Communications Lessons from the Frontline of Desalination Plant Development. Presentation given to the NRC Committee on Advancing Desalination Technology, Clearwater, FL.

Sallangos, O. L. V. 2004. Operating experience of the Dhekelia seawater desalination plnat using an innovative energy recovery system. Morocco: European Desalination Society.

Sallangos, O. L. V., and E. A. Kantilaftis. 2003. Operating Eexperience of the Dhekelia seawater desalination plant using an innovative energy recovery system. Presentation at the European Desalination Society Conference, Morocco, 2003.

San Francisco Bay Conservation and Development Commission. 2005. Desalination and San Francisco Bay. San Francisco, CA: San Francisco Bay Conservation and Development Commission.

Sarma, S. S., L. Beladjal, S. Nandini, G. Cerón-Martínez, and K. Tavera-Briseño. 2005. Effect of salinity stress on the life history variables of *Branchipus schaefferi* Fisher, 1834 (Crustacea: Anostraca). Saline Systems 1: 4.

Seacord, T. F. 2006a. Affordable Desalination Collaboration 2005. Presentation at AMTA Bi-Annual Conference, Anaheim, CA, July 2006.

Seacord, T. F. 2006b. Developing Post-Treatment Criteria for Seawater Reverse Osmosis. IDA Desalination Yearbook 2006-2007.

Seattle Public Utilities. 2007. Seattle Water Supply System Regional 1% Water Conservation Program. Saving Water Partnership 2006 Annual Report. Available online at: http://www.seattle.gov/util/stellent/groups/public/@spu/@csb/documents/webcontent/spu01_002837.pdf. Accessed March 14, 2008.

Seawater Desalination and the California Coastal Act. 2004. Chapter 5.1.3, p. 83.

Sedlak, D. L., and K. E. Pinkston. 2001. Factors Affecting the Concentrations of Pharmaceuticals Released to the Aquatic Environment. Proceedings of the National Groundwater Association, 2nd International Conference on Pharmaceuticals and Endocrine Disrupting Chemicals in Water, Minneapolis, MN, October 9-11. Westerville, OH: National Groundwater Association.

Sethi, S., S. Walker, P. Xu, and J. Drewes. 2006a. Existing and emerging concentrate minimization and disposal practices for membrane systems. Florida Water Resources Journal, June:38-48.

Sethi, S., S. Walker, P. Xu, and J. Drewes. 2006b. Desalination Product Water Recovery and Concentrate Volume Minimization, Phase I draft final report, AwwaRF Project No. 3030. Carollo Engineers and Colorado School of Mines.

Sethi, S., G. Juby, S. Li, A. Zacheis, T. Mulvihill, B. Standard, R. Morquecho, C. Krieger, and D. Scriven. 2007. Inland Brackish Water Desalination: Feasibility Study Evaluates Alternatives and Develops a Zero-Liquid-Discharge Solution. Proceedings of the AWWA Membrane Technology Conference, Tampa, FL, March 2007.

Sethi, S., P. Xu, and J. Drewes. 2008. Desalination Product Water Recovery and Concentrate Volume Minimization, Final Draft Report. AWWARF Project No. 3030. Carollo Engineers and Colorado School of Mines.

Shams El Din, A., and B. Makkawi. 1998. Operation processes affection corrosion in MSF distillers. Desalination 115:33-37.Sherwood, T. K., P. L. T. Brian, R. E. Fisher, and L. Dressner. 1964. Industrial Engineering Chemical Fundamentals 3:210.

Shih, H., and T. Shih. 2007. Utilization of waste heat in the desalination process. Desalination 204:464-470.

Shiklomanov, I. A. 1993. World fresh water resources. Pp. 13-24 in Water in Crisis: A Guide to the World's Fresh Water Resources, P. Gleick, ed. New York: Oxford University Press.

Simon, P. 1998. Tapped Out: The Coming World Crisis in Water and What We Can Do About It. New York: Welcome Rain Publishers.

Singer, P. C. 1999. Formation and Control of Disinfection Byproducts in Drinking Water. Denver, CO: AWWA.

Snyder, E. M, S. A. Snyder, J. P. Giesy, S. A. Blonde, G. K. Hurlburt, C. L. Summer, R. R. Mitchell, and D. M. Bush. 2000. SCRAM: A scoring and ranking system for persistent, bioaccumulative, and toxic substances for the North American Great Lakes—Part II Bioaccumulation potential and persistence. Environmental Science and Pollution Research 7:116-121.

Snyder, S. A., S. Adham, A. M. Redding, F. S. Cannon, J. DeCarolis, J. Oppenheimer, E. C. Wert, and Y. Yoon. 2007. Role of membranes and activated carbon in the removal of endocrine discruptors and pharmaceuticals. Desalination 202:156-181.

Solley, W. B., R. R. Pierce, and H. A. Perlman. 1998. Estimated Use of Water in the United States. U.S. Geological Survey Circular 1200. Reston, VA: USGS.

Song, L., J. Y. Hu, S. L. Ong, W. J. Ng, M. Elimelech, and M. Wilf. 2003a. Emergence of thermodynamic restriction and its implications for the full-scale reverse osmosis process. Desalination 155:213-228.

Song, L., J. Y. Hu, S. L. Ong, W. J. Ng, M. Elimelech, and M. Wilf. 2003b. Performance limitation in the full-scale reverse osmosis process. Journal of Membrane Science, 214:239-244.

Spiegler, K. S., and Y. M. El-Sayed. 1994. A Desalination Primer. L'Aquila, Italy: Balaban Desalination Publications.

Srisurichan, S., A. G. Fane, and R. Jiraratananon. 2006. Mass transfer mechanisms and transport resistances in direct contact membrane distillation processes. Journal of Membrane Science 277:186-194.

Standard & Poor. 2006. Global Project Finance Yearbook. New York, NY: McGraw Hill Companies, Inc.

Stoughton, R. W., and M. H. Lietzke. 1965. Calculation of some thermodynamic properties of sea and salt solutions at elevated temperatures from data of NaCl solutions. Journal of Chemical and Engineering Data 10:254-260.

Stratus Consulting. 2006. Institutional Issues in Desalination and Water Purification Technologies, Final Report. Prepared for Joint Water Reuse and Desalination Task Force, AWWA Research Foundation, Sandia National Laboratory, U.S. Bureau of Reclamation, WaterReuse Foundation. Boulder, CO: Stratus Consulting Inc.

Stumm, W., and J. J. Morgan. 1996. Aquatic Chemistry, 3rd edition. New York: John Wiley and Sons.

Summers, R. S., M. A. Benz, H. M. Shukairy, and L. Cummings. 1994. Effects of separation processes on the formation of brominated trihalomethanes. Journal of AWWA 86:88.

SWAQ (Subcommittee on Water Availability and Quality). 2007. A Strategy for Federal Science and Technology to Support Water Availability and Quality in the United States. Washington, D.C.: National Science and Technology Council.

SWRCB (California State Water Resources Control Board). 1995. Order No. 95-10. Sacramento, CA: SWRCB.

Tampa Bay Water. 2006. Tampa Bay Seawater Desalination Facility Hydrobiological Monitoring Summary 2002 through 2005. Tampa Bay, FL: PBS&J.

Taniguchi, Y., K. Ohta, T. Okabe, M. Hirai, and T. Goto. 1995. Pretreatment with membranes for sea water reverse osmosis process. Proceedings of the International Desalination Association 2:135-146.

Taylor, J..S., J. D. Dietz, S. K. Hong, A. A. Randall, and C. Owen. et al. 2005. The Effects of Blending on Distribution System Water Quality. AWWA Research Foundation.

Tergazi, K., and R. B. Peck. 1967. Soil Mechanics in Engineering Practice, 2nd edition. New York: John Wiley and Sons.

Texas Water Development Board. 2007. Texas water use from 1974-2004. Austin, TX: Texas Water Development Board. Available online at: http://www.twdb.state.tx.us/wushistorical/Desktop Default.aspx?PageID=1 Accessed November 16, 2007.

Thompson, R. G., M. L. Hyatt, J. W. McFarland, and H. P. Young. 1971. Forecasting Water Demands. Special Study No. 3 for the National Water Commission. Springfield, VA: National Technical Information Service.

Timmer, J. 2001. Properties of nanofiltration membranes; model development and industrial application. Doctoral thesis. Delft, The Netherlands: Eindhoven University of Technology.

Trieb, F., J. Nitsch, S. Kronshage, C. Schillings, L.-A. Brischke, G. Knies, and G. Czisch. 2003. Combined solar power and desalination plants for the Mediterranean region—sustainable energy supply using large-scale solar thermal power plants. Desalination 153:39-46.

Trussell, R. S., and R. R. Trussell. 2005. A Critical Review of Alternative Desalination Technologies. Prepared for the Joint Water Reuse & Desalination Task Force.

Tseng, T., R. Cheng, D. Vuong, and K. Wattier. 2003. Optimization of Dual-Staged Nanofiltration Membranes for Seawater

Desalination. Proceedings of the AWWA 2003 CA-NV Annual Fall Conference, October 2003.
Tzen, E., and R. Morris. 2003. Renewable energy sources for desalination. Solar Energy 75:375-379.
United Nations. 1987. Report of the World Commission on Environment and Development. General Assembly Resolution 42/187, December 11, 1987.
United Nations Intergovernmental Panel on Climate Change. 2007. Climate Change 2007: The Physical Science Basis, Summary for Policy Makers.
USAID (U.S. Agency for International Development). 1980. The USAID Desalination Manual. Washington, D.C.: CH2M HILL International for the U.S. Agency for International Development.
USBR (U.S. Bureau of Reclamation). 2003. Desalting Handbook for Planners, 3rd Edition. Washington, D.C.: USBR.
USBR. 2004. Solar and waste heat desalination by membrane distillation. Desalination and Water Purification Research and Development Program Report No. 81. Denver, CO: USBR.
USBR and Sandia National Laboratories. 2003. Desalination and Water Purification Technology Roadmap: A Report of the Executive Committee. Desalination & Water Purification Research & Development Report 95. Denver, CO: U.S. Department of the Interior, Bureau of Reclamation, Water Treatment and Engineering Group.
U.S. Census Bureau. 2001. Historical National Population Estimates: July 1, 1900 to July 1, 1999 (release date: September 27, 2000) and Resident Population Estimates of the United States by Age and Sex, April 1, 1990 to July 1, 1999, with Short-Term Projection to November 1, 2000 (release date: January 2, 2001). Washington, D.C.: Department of Commerce, Bureau of the Census.
U.S. Census Bureau. 2007. Statistical Abstract of the United States: 2007. Washington, D.C.: U.S. Department of Commerce, Bureau of the Census. Available online at: http://www.census.gov/prod/2006pubs/07statab/pop.pdf. Accessed November 16, 2007.
USDA and USFS (U.S. Department of Agriculture and U.S. Forest Service). 1989. RPA Assessment of the Forest and Rangeland Situation in the United States, 1989. Washington, D.C.: U.S. Department of Agriculture and U.S. Forest Service.
U.S. Department of Energy. 2006. Energy Demands on Water Resources. Report to Congress on the Interdependency of Energy and Water. Washington, D.C.: U.S. Department of Energy.
U.S. Energy Information Administration. 2007a. Annual Energy Review. Available online at: http://www.eia.doe.gov/emeu/aer/pdf/pages/

sec8_39.pdf. Accessed April 8, 2008.
U.S. Energy Information Administration. 2007b. Annual Oil Market Chronology. Available online at: http://www.eia.doe.gov/emeu/cabs/AOMC/Overview.html. Accessed April 8, 2008.
USGS (U.S. Geological Survey). 2003. Desalination of Ground Water: Earth Science Perspectives. Office of Ground Water. Reston, VA: USGS. Available online at: http://pubs.usgs.gov/fs/fs075-03/. Accessed December 5, 2007.
USGS. 2006. National Stream Water Quality Network (NASQAN) Data: Colorado Basin 1996-2005. Washington, D.C.: USGS. Available online at: http://water.usgs.gov/nasqan/data/finaldata/nib.html. Accessed December 11, 2007.
USGS. 2007a. Saltwater Intrusion. Northborough, MA: USGS. Available online at: http://water.usgs.gov/ogw/gwrp/saltwater/salt.html. Accessed May 27, 2008.
USGS. 2007b. Subsidence and Fault Activation Related to Fluid Energy Production, Gulf Coast Basin Project. Washington, D.C.: USGS. Available online at: http://coastal.er.usgs.gov/gc-subsidence/. Accessed November 16, 2007.
U.S. Senate Select Committee on National Water Resources. 1961. Report of the Select Committee. Report No. 29. Washington, D.C.: U.S. Senate Select Committee on National Water Resources.
Vengosh, A., E. Weinthal, and W. Kloppman. 2004. Natural boron contamination in Meditteranean groundwater. Geotimes May 2004. Available online at: http://www.agiweb.org/geotimes/may04/feature_boron.html. Accessed March 14, 2008.
Viessman, W. 1980. Assessing the Nation's Water Resources: Issues and Options. Congressional Research Service of the Library of Congress: For the U.S. Senate Committee on Environment and Public Works. Serial No. 96-19, USGPO. Washington, D.C.: Congressional Research Service.
Voss, C. I., and A. M. Provost. 2003. SUTRA: A Model for Saturated-Unsaturated, Variable-Density Ground-Water Flow with Solute or Energy Transport. U.S. Geological Survey Water-Resources Investigations Report 02-4231. Reston, VA: USGS. Available online at: http://water.usgs.gov/nrp/gwsoftware. Accessed December 12, 2007.
Voutchkov, N. 2004. Bring down desalination costs by co-location. EDS Newsletter Issue 21, October.
Voutchkov, N. 2005. Shared infrastructure benefits desalination economics. Available online at: http://www.waterandwastewater.com/www_services/news_center/publish/article_00693.shtml. Accessed April 26, 2007.

Walz, T., B. L. Smith, M. L. Zeidel, A. Engel, and P. Agre. 1994. Biologically active two-dimensional crystals of aquaporin CHIP. Journal of Biological Chemistry 269(3):1583-1600.

Wangnick, K. 2002. 2002 IDA Worldwide Desalting Plants Inventory. Report No. 17. Gnarrenburg, Germany: Wangnick Consulting GMBH.

Water Corporation. 2007. The Perth Seawater Desalination Plant: Wind Farm to Provide Power. Available on line at: http://water.com.au/_files/Desalination_windpower.pdf. Accessed November 16, 2007.

Weast, R., M. Astle, and W. Beyer. 1985. CRC Handbook of Chemistry and Physics, 69th edition. Boca Raton, FL: CRC Press.

Weber, W. J. 1972. Physicochemical Processes for Water Quality Control. New York: John Wiley and Sons.

Weis, J. S., J. Skurnick, and P. Weis. 2004. Studies of a contaminated brackish marsh in the Hackensack Meadowlands of northeastern New Jersey: Benthic communities and metal contamination. Marine Pollution Bulletin 49:1025-1035.

Whitcomb, J. B. 2005. Florida Water Rate Evaluation of Single Family Homes. Brooksville, FL: Southwest Florida Water Management District. Available online at: http://www.swfwmd.state.fl.us/documents/reports/water_rate_report.pdf. Accessed on April 8, 2008.

WHO (World Health Organization). 2003. Total dissolved solids in drinking-water. Geneva, Switzerland: WHO. Available online at: http://www.who.int/water_sanitation_health/dwq/chemicals/tds.pdf. Accessed March 14, 2008.

WHO. 2004. Guidelines for drinking-water quality, Volume 1 Recommendations, 3rd edition. Geneva, Switzerland: WHO. Available online at: http://www.who.int/water_sanitation_health/dwq/gdwq3/en/. Accessed June 12, 2008.

WHO. 2007. Desalination for Safe Water Supply: Guidance for the Health and Environmental Aspects Applicable to Desalination. Geneva: WHO. Available online at: http://www.who.int/water_sanitation_health/gdwqrevision/desalination.pdf. Accessed January 4, 2008.

Williams, M. D., and M. Pirbazzi. 2003. Biological Sulfate Reduction (BSR) for Recovering Reverse Osmosis Brine. Quality Assurance Project Plan for No. 51539. Desalination Research and Innovation Partnership (DRIP).

Williams, M. D., R. Evangelista, and Y. Cohen. 2002. Non-thermal Process for Recovering Reverse Osmosis Concentrate: Process Chemistry and Kinetics. Proceedings of the 2002 AWWA Water Quality Technology Conference, Seattle, WA.

Winters, H. 2001. Identification of critical flux and cross flow conditions for the control of bacterial and organic fouling of seawater reverse osmosis membranes. MEDRC Project 97-AS-004b.

Winters, H., A.G. Fane, and B. Nicolaisen. 2007. Microbial Membrane Fouling: Microbial Cake Enhanced Osmotic Pressure (CEOP) or Thermodynamic Restriction? Presentation at IDA World Congress, Mas Palomas, Spain, October 2007.

Wollman, N., and G. W. Bonem. 1971. Outlook for Water: Quality, Quantity, and National Growth. Baltimore, MD: Resources for the Future, Johns Hopkins University Press.

Woodward Clyde Consultants. 1991. Final environmental impact report (FEIR). City of Santa Barbara and Ionics, Incorporated's Temporary Emergency Desalination Project, SB-106-90, State Clearing House No. 9010859.

World Commission on Dams. 2000. Dams and Development: A New Framework for Decision-Making. London: Earthscan.

WRC (Water Resource Council). 1968. The Nation's Water Resources Part 1, First National Water Assessment. Washington, D.C.: U.S. Government Printing Office (USGPO).

WRC. 1978. The Nation's Water Resources 1975-2000. Washington, D.C.: USGPO.

Xu, J., G. Ruan, X. Chu, Y. Yao, B. Su, and C. Gao. 2007. A pilot study of UF pretreatment without any chemicals for SWRO desalination in China. Desalination 207:216-226.

Xu, T. 2005. Ion exchange membranes: State of their development and perspective. Journal of Membrane Science 263:1-29.

Xu, Y., Y. Y. Xu, and B. K. Zhu. 2006. Pilot test of vacuum membrane distillation for seawater desalination on a ship. Desalination 189:165-169.

Yallaly, B., T. Seacord, and T. Kalkman. 2007. Meeting Criteria for Water and Wastewater Systems Simultaneously Using Reverse Osmosis and Zero Liquid Discharge Technology. Ridgecrest, CA: Indian Wells Valley Water District.

Yermiyahu, U., A. Tal, A. Ben-Gal, A. Bar-Tal, J. Tarchitzky, and O. Lahav. 2007. Rethinking desalinated water quality and agriculture. Science 318:920-921.

York, R., and M. Foster. 2005. Issues and environmental impacts associated with once-through cooling at California's coastal power plants. Sacramento, CA: California Energy Commission. Available online at: http://www.energy.ca.gov/2005publications/CEC-700-2005-013/CEC-700-2005-013.PDF. Accessed February 27, 2008.

Younos, T. 2005. Environmental issues of desalination. University Council on Water Resources Journal of Contemporary Water Research and Education 132:11-18.

Zhang, J., S. Gao, H. Zeng, F. Zhang, C. Li, Y. Liu, D. Fu, and C. Ye. 2006. Pilot testing of two inside-out UF modules prior to RO for highturbidity seawater desalination. Desalination 196:66-75.

Glossary

Acre-foot (AF)—The volume of water that would cover a one acre area one foot deep. Equivalent to approximately 1,233.6 cubic meters or 325,900 gallons.
Applied research—Systematic study to gain knowledge or understanding necessary to determine the means by which a recognized and specific need may be met.
Available energy—Mechanical, electrical energy, or any other energy, which in practice can be nearly completely converted into mechanical work.

Basic research—Systematic study directed toward fuller knowledge or understanding of the fundamental aspects of phenomena and of observable facts without specific applications towards processes or products in mind.
Biofouling—Presence and growth of organic matter in a water treatment system that interferes with system performance.
Brackish water—Water with a salinity between that of fresh- and seawater.
Brine—Water with a greater salinity than seawater, usually in excess of 37,000 mg/L.

Capacitive deionization—An electrosorption process whereby ions are removed from water using an electric field gradient as the driving force.
Coagulation—A pretreatment process used in some desalination plants. A substance (e.g., ferric chloride) is added to a solution to cause suspended particles to agglomerate and form larger particles which are easier to remove from a solution than small particles.
Cogeneration—Dual-purpose facilities that produce both electricity and water. Cogeneration plants integrate the thermal desalination process with available steam from power generation.

Colloid—Suspended solid with a diameter less than 1 micron that can not be removed by sedimentation alone.
Concentrate—The water containing the dissolved solids removed during desalination.
Concentrate management—The handling and disposal or reuse of waste residuals from the desalination system.
Concentration polarization—A phenomenon in which solutes form a dense, polarized layer next to a membrane surface which eventually restricts flow through the membrane.
Consumptive use—Water withdrawn from a source and made unavailable for reuse in the same basin, such as through conversion to steam, losses to evaporation or transpiration, seepage to a saline sink, or contamination. Also referred to as irretrievable or irrecoverable loss.

Deaeration—Removal of oxygen. A pretreatment process in desalination plants to reduce corrosion.
Demand—An economic concept that is used to describe a want for water backed up by a willingness to pay.
Demand schedule (or curve)—A summarization of the quantities and qualities of water that consumers are willing to take at different prices.
Desalination—The process that removes dissolved solids, primarily salts and other inorganic constituents, from a saline water source.
Design-Build (DB)—A delivery approach characterized by a single contractual relationship between the public water provider and a contractor, who develops the project design and oversees its construction. The project is then operated by the owner or contract operator.
Design-Build-Operate (DBO)—A delivery approach involving a single contractor for design, construction, and operation.
Design-Build-Own-Operate-Transfer (DBOOT)—An expansion of the DBO concept in which the contractor also finances the project and initially owns the facility.
Development—Systematic application of knowledge or understanding, directed toward the production of useful materials, devices, and systems or methods, including design, development, and improvement of prototypes and new processes to meet specific requirements.
Dewvaporation—A desalination method where a stream of air is humidified by a falling film of saline water along one side of a heat transfer surface. The air is partially heated by an external source (e.g., solar, waste heat). The heated air then is swept along the condensing side of heat transfer films, where the vapor condenses to a liquid, which is collected as product water
Diffusion—The movement of suspended or dissolved particles or molecules from a more concentrated to a less concentrated area.

Distillation—A process of desalination where the intake water is heated to produce steam. The steam is then condensed to produce product water with extremely low salt concentration.
Drinking water—Water safe for human consumption or which may be used in the preparation of food or beverages, or for cleaning articles used in the preparation of food or beverages.

Effluent—Water leaving a desalting process. May be applied to both concentrate or product water.
Electrodialysis (ED)—A process of desalination whereby an electrical current is used to separate out salt and impurities in the intake water through the use of semipermeable, ion-selective membranes operating in a DC electric field.
Electrodialysis reversal (EDR)—A variation of the electrodialysis process using electrode polarity reversal to reduce and minimize scaling and fouling, thus allowing the system to operate at comparatively higher recoveries.
Entrainment—The incorporation of small organisms, including the eggs and larvae of fish and shellfish, into an intake system.
Evaporation—The process by which water is converted to a vapor.
Evaporation ponds—A concentrate management method where the concentrate is pumped into a shallow lined pond and allowed to evaporate naturally using solar energy.
Evaporation rate—The mass quantity of water evaporated from a specified water surface per unit of time.

Feedwater—Input or raw water stream fed into the desalination process.
Flux—Term used in reverse osmosis to indicate the rate of water permeation through the membrane. Usually expressed as gallons per square foot per day in the U.S., and liters per square meter per hour in metric units.
Forward osmosis—A membrane-based separation process which uses osmotic pressure difference between a concentrated "draw" solution and a feed stream to drive water through a semi-permeable membrane.
Fouling—The reduction in performance of process equipment (heat transfer tubing, membranes, etc.) that occurs as a result of scale buildup, biological growth, or the deposition of colloidal material.
Freeze desalination—Production of distillate by freezing a saline solution and separating the salt water from the pure crystalline water prior to melting.
Fresh water—Water that generally contains less than 1,000 mg/L of dissolved solids.

Gained output ratio (GOR)—A measure of the efficiency of a thermal desalination process, expressed as a ratio of the mass of water produced by the process over the mass of (saturated) steam supplying process heat. If, for example, 8 kg of water are produced from the desalination system for each kg of steam delivered to the process, the GOR is equal to 8. This measure is a dimensionless value.

Hybrid—A system incorporating multiple processes or technologies; for example a desalination facility incorporating both thermal and membrane processes. Generally the technologies should at least be partially integrated for some process benefit to qualify as hybrid.

Impingement—The pinning and trapping of fish or other larger organisms against the screens of water intake structures.
Intakes—The structures used to extract source water and convey it to the process system.
Ion exchange—A chemical process involving the reversible exchange of ions between a liquid and a solid.

Land subsidence—A gradual settling or sudden sinking of the Earth's surface owing to subsurface movement of earth materials, sometimes caused by groundwater withdrawal.
Louvers—A series of vertical panels placed perpendicular to the intake approach flow that serve to create a new velocity field that carries fish away from the intake and towards a fish bypass system.

Mechanical vapor compression (MVC)—See "vapor compression evaporation."
Membrane— In desalination, used to describe a semipermeable film. Membranes used in electrodialysis are permeable to ions of either positive or negative charge. Reverse osmosis and nanofiltration membranes ideally allow the passage of pure water and allow only minor passage of salts.
Microfiltration (MF)—Membranes used to reduce turbidity and remove suspended particles, algae, and bacteria. MF membranes operate via a sieving mechanism under a lower pressure than either UF or NF membranes, through membrane pores of 0.03 to 10 μm and an molecular weight cut off of greater than 100,000 daltons.
Multiple effect distillation (MED)—A thin film evaporation process where the vapor formed in a chamber, or effect, condenses in the next, providing a heat source for further evaporation.
Multi-port diffusion—The employment of multiple outlet ports, rather than a single pipe, during the process of diffusion, thus allowing the mix-

ing and dilution of the concentrate to be accelerated and lessening potential impacts in sensitive areas.
Multi-stage flash distillation (MSF)—A desalination process where a stream of brine flows through the bottom of chambers, or stages, each operating at a successively lower pressure, and a proportion of it flashes into steam and is then condensed.

Nanofiltration (NF)—Membranes used for water softening, organics and sulfate removal, and some removal of viruses. Pressure-driven removal is by combined particle size-based sieving and solution/diffusion. Pores in NF membranes are usually smaller than 0.001 μm and a molecular weight cutoff of 1,000 to 10,000 daltons.
Need—The minimum amount of water required to satisfy a particular purpose or requirement.

Passive screen—An intake screen that can be flushed back with compressed air, has no moving parts, and operates with a very low velocity (to mitigate impingement).
Performance ratio (PR)—The ratio of the mass of water produced by a desalination process over a fixed quantity of energy consumed. This ratio is useful, as steam may be delivered over a wide range of temperatures, and its heat content, both in terms of total enthalpy and in enthalpy of vaporization, differs at each temperature. Additionally, in many practical cases, steam may not be the medium of heat transfer. The Performance Ratio is most commonly defined as the mass, in pounds, of water produced by desalination per 1000 BTU of heat provided to the process. The SI equivalent of this formulation is the number of kg of water produced per 2326 kJ of heat.
Permeate—The liquid that passes through a membrane.
Post-treatment—The addition of chemicals to the product water to prevent corrosion of downstream infrastructure piping.
Precipitate—A substance separated from a solution by chemical or physical change as an insoluble amorphous or crystalline solid.
Pre-treatment—Refers to methods for treatment of water to remove suspended particles and control of biological growth, and to prepare the source water for further processing. Conventionally this involves coagulation, settling, and filtration. More recently ultrafiltration or microfiltration can be used to prepare the water.
Product water—Water produced as a result of treatment or desalination processes.

Recovery—In reverse osmosis processes, recovery indicates the amount/percentage of product water recovered from the feed stream.

Reverse osmosis (RO)—A method of separating water from dissolved salts by passing feedwater through a semipermeable membrane at a pressure greater than the osmotic pressure caused by the dissolved salts. RO operates via a solution/diffusion mechanism whereby the water dissolves into and diffuses through the nonporous membrane, leaving the majority of the salts behind in the concentrate. RO membranes are also capable of removing some larger organic contaminants. Small uncharged species can pass through the membrane. Water that passes through the membrane leaves the unit as permeate or product water; most of the dissolved impurities remain behind and are discharged in a brine or waste stream.
Ristroph screen—A modified traveling screen with water-filled lifting buckets that collect impinged organisms and transport them to a bypass, trough, or other protected area.

Saline water—Water with dissolved solids exceeding the limits of potability. Saline water may include sea water, brackish water, mineralized ground and surface water, and irrigation return flows.
Salinity—The concentration of dissolved salts in water.
Scaling—Mineral deposits or precipitates that form on the interior surfaces of process equipment or water lines as a result of heating or other physical or chemical change.
Sedimentation—The removal of settleable suspended solids from water or wastewater by gravity in a quiescent basin or clarifer.
Silt Density Index (SDI)—A measure of the fouling tendency of water, based on the timed flow of a liquid through a membrane filter at a constant pressure.
Solubility—A measure of the maximum amount of a certain substance that can dissolve in a given amount of water, or other solvent, at a given temperature.
Surface intakes—Often called open intakes, they are in direct communication with the ocean or sea where feedwater can be taken directly from the surface or below the surface through submerged intakes.
Subsurface intakes—Where feedwater is taken from below the floor of the ocean using natural occurring sand and geologic formations to provide filtration. Subsurface intakes can be horizontally drilled from central wells, slant-drilled from onshore beaches, or constructed infiltration beds.

Thermal vapor compression (TVC)—See "vapor compression evaporation".
Total dissolved solids (TDS)—The mass of all inorganic and organic material per unit volume of water after a sample has been filtered to remove suspended solids.

Ultrafiltration (UF)—Membrane removal of high-weight dissolved organic compounds, bacteria, and some viruses. UF membranes operate via a pressure-driven size-based sieving mechanism through a membrane with pores in the range of 0.002 to 0.1 μm with an molecular weight cut off of 10,000 to 100,000 daltons.

Vapor compression evaporation (VC/VCE)—Thermal desalination processes that utilize heat from compression of water vapor for subsequent evaporation of feedwater, either with a mechanical compressor (mechanical vapor compression, MVC) or a steam ejector (thermal vapor compression, TVC). Vapor compression processes are particularly useful for small to medium installations.

Water reclamation—The restoration of wastewater to a state that will allow its beneficial reuse.

Water reuse—The beneficial use of reclaimed water, such as for irrigation, cooling or washing.

Withdrawal—Water removed from a source and used to meet a human need.

Zero liquid discharge (ZLD)—A concentrate management approach in which no liquid effluent is discharged beyond the facility boundaries.

Appendix A

Desalination Federal Funding Survey

National Academy of Sciences
Committee on Advancing Desalination Technology
Federal Funding Survey
February 15, 2007

Background

As water supplies face growing demands, desalination and water reuse continue to evolve as an alternative to expand usable water supplies. Desalination technologies are no longer applied only to seawater source waters. Applications have expanded to convert inland brackish ground- and surface waters in many Western states over the past 5-10 years. A comparison by Reclamation indicates that the cost to convert saline and other impaired sources of water into usable water supplies is now competitive with, or lower than, many of the water rates charged throughout the United States and abroad. This suggests that the long-sought research goal of lowering the cost of desalination technologies so that it is a more cost-competitive water supply alternative may now be achieved or within reach.

Against this backdrop, the commercial sector, nonprofit organizations, academia, international entities, and multiple federal, state, and local agencies have been investing more time and money in implementing, maturing, and advancing desalination technologies. Because of these recent advances, and increased national and international attention, the future directions for advancing and implementing desalination and water purification technologies are at a critical juncture in the United States. Reclamation and the Environmental Protection Agency are currently funding the National Academy of Sciences (NAS) to study the role of the federal government in advancing desalination technology.

Assessing the current situation also indicates that a strong, international industry presence has emerged around desalination and other water purification technologies. When there is a strong industry presence, market forces have the ability to advance and mature technologies through industry investments. Under these conditions, the need and role for federal research investments should be carefully reassessed. Reclamation has requested definition of the appropriate role for federal research as one of the key tasks for the NAS Committee on Advancing Desalination Technology to address, as described in the Project Scope below.

As part of the President's Management Agenda, the Administration's research and development (R&D) investment criteria direct federal agencies to prevent federal R&D investments in areas with established industries where the federal investment may discourage, or even displace, industry investments that would otherwise occur. Because Reclamation views the state of technology and the future role of research to be at a critical juncture, the first step in Reclamation's strategy is having the NAS conduct a thorough evaluation of the state of technology, opportunities, obstacles, and the practical aspects of implementing these technologies.

National Academy of Science Committee on Advancing Desalination Technology
R&D Funding Survey

In order to better understand the current scope of the federal role in desalination research, the NAS committee would appreciate receiving some input from the agencies involved in desalination research and development.

Agency:

Contact:
Title:
Phone:
Email:

1. Please provide budget information for your agencies desalination R&D (basic research, applied research, and development – see OMB definitions at the bottom of the next page*) for FY2005, FY2006, and FY2007. Please note that subsidies for the construction of desalination facilities should NOT be included here but will be addressed in question #8.

	FY2005	FY2006	FY2007
Requested			
Appropriated			
Expended			

2. Please provide topical categories of desalination research that your agency conducts. Please provide project titles (at the minimum) for currently supported research.

3. Please approximate the percentage of your investment in basic research, applied research, and development (following the OMB definitions below*).

4. What percentage of your investment do you consider to be in high-risk versus low-risk research areas? Please provide your definition of “high-risk” research.

5. What percentage of your investment is in long-term research (> 3 years for government) versus short-term research (< 3 years for government)?

6. Irrespective of your organization's mission, what do you think are the nation's top high-risk desalination research priorities?

7. What do you think are the most promising technologies or innovations for advancing desalination?

8. Does your agency provide funds to subsidize the construction of desalination facilities for coastal seawater, inland brackish groundwater, or inland surface water? If so please provide budget information in the table below.

	FY2005	FY2006	FY2007
Requested			
Appropriated			
Expended			

9. Does your agency fund advanced water treatment research for water reuse and/or water recycling applications (e.g., reverse osmosis)? If so please provide budget information for advanced water treatment research not included in your answer to question #1 in the table below.

	FY2005	FY2006	FY2007
Requested			
Appropriated			
Expended			

10. Please briefly describe your agency's historical involvement in desalination research prior to FY2005. Please provide levels of investment and types of research.

*OMB Research Definitions (OMB Circular A-11)

Basic Research – "Basic research is defined as systematic study directed toward fuller knowledge or understanding of the fundamental aspects of phenomena and of observable facts without specific applications towards processes or products in mind."

Applied Research – "Applied research is defined as systematic study to gain knowledge or understanding necessary to determine the means by which a recognized and specific need may be met."

Development – "Development is defined as systematic application of knowledge or understanding, directed toward the production of useful materials, devices, and systems or methods, including design, development, and improvement of prototypes and new processes to meet specific requirements."

Appendix B

Mass and Energy Balance on Reverse Osmosis System

Mass and Energy Balance on Reverse Osmosis System:

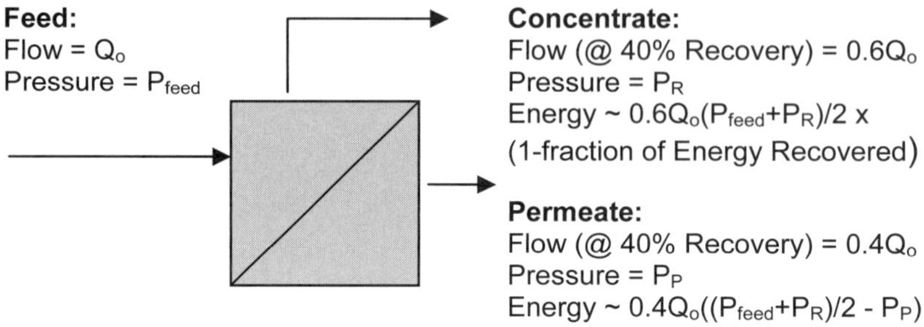

Following is an approximation of the energy used in a typical RO process operating at 40% recovery and an energy recovery device operating at an efficiency of η_{eff}.

$$\text{Energy Used} \cong 0.6\text{Q}_\text{O}\frac{(\text{P}_\text{O}+\text{P}_\text{R})}{2}(1-\eta_{\text{eff}})+0.4\text{Q}_\text{O}(\frac{\text{P}_\text{O}+\text{P}_\text{R}}{2}-\text{P}_\text{P}).$$

Making the assumption that P_p is significantly less than the applied average operating pressure,

(i.e., that $\text{P}_\text{P} = 14.7$ at atmospheric pressure and

because this term is $<< \frac{\text{P}_0+\text{P}_\text{R}}{2}$ this term is presumed to be negligible and $\cong 0$)

and taking a ratio of a future, new energy balance based on a new membrane with new properties relative to a baseline energy balance we get the following equation:

$$\frac{\text{Energy}_{\text{New}}}{\text{Energy}_{\text{Baseline}}} = \frac{0.6\text{Q}_\text{O}\frac{(\text{P}_{\text{ON}}+\text{P}_{\text{RN}})}{2}(1-\eta_{\text{eff}})+0.4\text{Q}_O\frac{(\text{P}_{\text{ON}}+\text{P}_{\text{RN}})}{2}}{0.6\text{Q}_\text{O}\frac{(\text{P}_\text{O}+\text{P}_\text{R})}{2}(1-\eta_{\text{eff}})+0.4\text{Q}_O\frac{(\text{P}_\text{O}+\text{P}_\text{R})}{2}}$$

This equation can be factored as follows,

$$\frac{\text{Energy}_{\text{New}}}{\text{Energy}_{\text{Baseline}}} = \frac{\frac{(P_{ON} + P_{RN})}{2}(0.6Q_O(1-\eta_{\text{eff}}) + 0.4Q_O)}{\frac{(P_O + P_R)}{2}(0.6Q_O(1-\eta_{\text{eff}}) + 0.4Q_O)}$$

and further simplified to the following equation:

$$\frac{\text{Energy}_{\text{New}}}{\text{Energy}_{\text{Baseline}}} = \frac{\frac{(P_{ON} + P_{RN})}{2}}{\frac{(P_O + P_R)}{2}} = \frac{P_{\text{Avg. Applied New}}}{P_{\text{Avg. Applied Baseline}}} = \frac{P_{\text{Avg. Driving New}} + P_{\text{Osmotic}}}{P_{\text{Avg. Applied Baseline}} + P_{\text{Osmotic}}}$$

.

As shown above, the average applied pressure can be broken down into two components: (1) the osmotic pressure required to overcome the osmotic energy barrier and (2) the net driving pressure required to overcome the native resistance of the membrane permeability.

For the purposes of illustrating the sensitivity of membrane permeability on potential future energy reductions, the following system operating data is taken from "The Guidebook to Membrane Desalination Technology," p. 472, Balaban Desalination Publications, 2007:

Average Total Dissolved Solids (TDS): 59,921 ppm

Temperature: 26 °C

$P_{\text{Avg Osmotic}}$ = 656 psi (at the average TDS of 59,053 ppm and Temperature of 28 °C; calculated by using the Van't Hoff equation)

$P_{\text{Avg Applied Baseline}}$ = 936 psi (P_{feed} = 947 psi; $P_{\text{concentrate}}$ = 924 psi)

$P_{\text{Avg Driving Baseline}}$ = (936 – 656) = 280 psi

$P_{\text{Avg Driving New}}$ = 0.5 x 280 = 140 psi (Reflecting a doubling of membrane permeability)

Assuming that the membrane permeability can be doubled without sacrificing salt rejection, the average driving pressure for the new membrane can be reduced by 50 percent (shown above).

Substituting these values into the equation above,

$$\frac{\text{Energy}_{\text{New}}}{\text{Energy}_{\text{Baseline}}} = \frac{P_{\text{Avg. Driving New}} + P_{\text{Osmotic}}}{P_{\text{Avg. Applied Baseline}} + P_{\text{Osmotic}}} = \frac{140 + 656}{280 + 656} = 0.85$$

results in an energy ratio of new to baseline of 0.85. This translates into a net reduction of energy equal to 15 percent from today's baseline.

Appendix C

Desalination Economics Summary Data

CapEx Scale Sensitivity & OpEx Scale and Interest Rate Sensitivity Analysis

CAPEX ($/m^3)	38 km^3/day	189 km^3/day	380 km^3/day
Process Unit Operation Costs			
Conventional pretreatment	$133	$115	$106
UF/MF pretreatment	$150	$130	$120
First-pass SW RO	$1,239	$1,069	$989
Second-pass SW RO (optional)	$140	$122	$113
Total Process Costs			
Conventional Pretreatment & 2-Stage RO Process ($/m^3)	$1,512	$1,306	$1,208
OpEx on capital			
25-yr depreciation; 3% cost of capital	$0.24	$0.20	$0.19
25-yr depreciation; 5% cost of capital	$0.29	$0.25	$0.23
25-yr depreciation; 7% cost of capital	$0.35	$0.30	$0.28
Conventional Pretreatment & 2-Stage BW RO Process ($/m^3)	$413	$359	$332
OpEx on capital			
25-yr depreciation; 3% cost of capital	$0.06	$0.06	$0.05
25-yr depreciation; 5% cost of capital	$0.08	$0.07	$0.06
25-yr depreciation; 7% cost of capital	$0.10	$0.08	$0.08
UF/MF Pretreatment & 2-Stage RO Process ($/m^3)	$1,529	$1,321	$1,222
OpEx on capital			
25-yr depreciation; 3% cost of capital	$0.24	$0.21	$0.19
25-yr depreciation; 5% cost of capital	$0.29	$0.25	$0.24
25-yr depreciation; 7% cost of capital	$0.36	$0.31	$0.28

OpEx - Scale Sensitivity Analysis

Membrane Life 5 yrs; Baseline Operating Pressure; $0.07/kWh; 5% Cost of $$$

Scale:	38 km^3/d	189 km^3/d	380 km^3/d
2 Stage SW RO OpEx on Capital - Conventional Pretreatment 25-yr depreciation; 5% cost of capital	$0.29	$0.25	$0.23
2 Stage SW RO OpEx on Capital - UF/MF Pretreatment 25-yr depreciation; 5% cost of capital	$0.29	$0.25	$0.24
2 Stage BW RO OpEx on Capital - Conventional Pretreatment 25-yr depreciation; 5% cost of capital	$0.08	$0.07	$0.06
Conventional Pretreatment	$0.07	$0.07	$0.07
Chemicals ($/$m^3$)	$0.04	$0.04	$0.04
Filters ($/$m^3$)	$0.00	$0.00	$0.00
Energy ($/$m^3$)	$0.03	$0.03	$0.03
UF/MF Pretreatment	$0.03	$0.03	$0.03
Chemicals ($/$m^3$)	$0.01	$0.01	$0.01
Membranes ($/$m^3$)	$0.01	$0.01	$0.01
Energy ($/$m^3$)	$0.01	$0.01	$0.01
First Pass SW RO	$0.23	$0.23	$0.23
Chemicals (antiscalant mostly, $/$m^3$)	$0.02	$0.02	$0.02
Membranes (5-yr life, $/$m^3$)	$0.02	$0.02	$0.02
Energy ($/$m^3$)	$0.19	$0.19	$0.19
Second Pass SW RO (optional)	$0.05	$0.05	$0.05
Chemicals (NaOH mostly, $/$m^3$)	$0.01	$0.01	$0.01
Membranes (5-yr life, $/$m^3$)	$0.01	$0.01	$0.01
Energy ($/$m^3$)	$0.03	$0.03	$0.03
Permeate Conditioning	$0.03	$0.03	$0.03
Chemicals ($/$m^3$)	$0.02	$0.02	$0.02
Energy ($/$m^3$)	$0.01	$0.01	$0.01
Total Labor	$0.03	$0.03	$0.03
Total Maintenance	$0.04	$0.04	$0.04
Total OpEx Costs - 2-Stage SW RO Conventional PreTreatment ($/1000 gallons) =	$0.74	$0.70	$0.68
Total OpEx Costs - 2-Stage SW RO UF/MF PreTreatment ($/1000 gallons) =	$0.70	$0.66	$0.65
Total OpEx Costs - 2-Stage BW RO Conventional PreTreatment ($/1000 gallons) =	$0.30	$0.29	$0.28

OpEx - Membrane Life Cost Sensitivity Analysis

Baseline Operating Pressure; $0.07/kWh; 189 km^3/d; 5% Cost of Capital

Membrane Life:	0.5 yr	1 yr	2 yr	3 yr	5 yr	10 yr
OpEx on Capital - Conventional Pretreatment 25-yr depreciation; 5% cost of capital	$0.25	$0.25	$0.25	$0.25	$0.25	$0.25
OpEx on Capital - UF/MF Pretreatment 25-yr depreciation; 5% cost of capital	$0.25	$0.25	$0.25	$0.25	$0.25	$0.25
Conventional Pretreatment	$0.07	$0.07	$0.07	$0.07	$0.07	$0.07
Chemicals ($/m^3)	0.04	0.04	0.04	$0.04	$0.04	$0.04
Filters ($/m^3)	0.00	0.00	0.00	$0.00	$0.00	$0.00
Energy ($/m^3)	0.03	0.03	0.03	$0.03	$0.03	$0.03
UF/MF Pretreatment	$0.00	$0.00	$0.00	$0.04	$0.03	$0.03
Chemicals ($/m^3)				$0.01	$0.01	$0.01
Membranes ($/m^3)				$0.02	$0.01	$0.00
Energy ($/m^3)				$0.01	$0.01	$0.01
First-Pass SW RO	$0.39	$0.30	$0.25	$0.24	$0.22	$0.22
Chemicals (antiscalant mostly, $/m^3)	0.02	0.02	0.02	$0.02	$0.02	$0.02
Membranes ($/m^3)	$0.18	$0.09	$0.04	$0.03	$0.02	$0.01
Energy ($/m^3)	0.19	0.19	0.19	$0.19	$0.19	$0.19
Second-Pass SW RO (optional)	$0.14	$0.09	$0.06	$0.05	$0.05	$0.04
Chemicals (NaOH mostly, $/m^3)	0.01	0.01	0.01	$0.01	$0.01	$0.01
Membranes ($/m^3)	$0.10	$0.05	$0.02	$0.02	$0.01	$0.00
Energy ($/m^3)	0.03	0.03	0.03	$0.03	$0.03	$0.03
Permeate Conditioning	$0.03	$0.03	$0.03	$0.02	$0.02	$0.02
Chemicals ($/m^3)	0.02	0.02	0.02	$0.02	$0.02	$0.02
Energy ($/m^3)	0.01	0.01	0.01	$0.01	$0.01	$0.01
Total Labor	0.03	0.03	0.03	$0.03	$0.03	$0.03
Total Maintenance	0.04	0.04	0.04	$0.04	$0.04	$0.04
Total OpEx Costs - Conventional Pretreatment ($/1000 gallons) =	$0.95	$0.81	$0.74	$0.70	$0.69	$0.68

OpEx - Applied Pressure Cost Sensitivity Analysis

Membrane Life 5 yrs; $0.07/kWh; 189 km^3/d; 5% Cost of Capital

Applied Pressure Sensitivity:	25% reduction	10% reduction	Base Case	25% Increase	50% Increase
OpEx on Capital - Conventional Pretreatment 25-yr depreciation; 5% cost of capital	$0.25	$0.25	$0.25	$0.25	$0.25
OpEx on Capital - UF/MF Pretreatment 25-yr depreciation; 5% cost of capital	$0.25	$0.25	$0.25	$0.25	$0.25
Conventional Pretreatment	$0.07	$0.07	$0.07	$0.07	$0.07
Chemicals ($/$m^3$)	$0.04	$0.04	$0.04	$0.04	$0.04
Filters ($/$m^3$)	$0.00	$0.00	$0.00	$0.00	$0.00
Energy ($/$m^3$)	$0.03	$0.03	$0.03	$0.03	$0.03
UF/MF Pretreatment	$0.03	$0.03	$0.03	$0.04	$0.04
Chemicals ($/$m^3$)	$0.01	$0.01	$0.01	$0.01	$0.01
Membranes ($/$m^3$)	$0.01	$0.01	$0.01	$0.01	$0.01
Energy ($/$m^3$)	$0.01	$0.01	$0.01	$0.02	$0.02
First-Pass SW RO	$0.20	$0.21	$0.22	$0.27	$0.32
Chemicals (antiscalant mostly, $/$m^3$)	$0.02	$0.02	$0.02	$0.02	$0.02
Membranes (5-yr life, $/$m^3$)	$0.02	$0.02	$0.02	$0.02	$0.02
Energy ($/$m^3$)	$0.17	$0.17	$0.19	$0.24	$0.28
Second-Pass SW RO (optional)	$0.05	$0.05	$0.05	$0.05	$0.05
Chemicals (NaOH mostly, $/$m^3$)	$0.01	$0.01	$0.01	$0.01	$0.01
Membranes (5-yr life, $/$m^3$)	$0.01	$0.01	$0.01	$0.01	$0.01
Energy ($/$m^3$)	$0.03	$0.03	$0.03	$0.03	$0.03
Permeate Conditioning	$0.02	$0.02	$0.02	$0.02	$0.02
Chemicals ($/$m^3$)	$0.02	$0.02	$0.02	$0.02	$0.02
Energy ($/$m^3$)	$0.01	$0.01	$0.01	$0.01	$0.01
Total Labor	$0.03	$0.03	$0.03	$0.03	$0.03
Total Maintenance	$0.04	$0.04	$0.04	$0.04	$0.04
Total OpEx Costs - Conventional Pretreatment ($/1000 gallons) =	$0.66	$0.67	$0.69	$0.74	$0.79
Total OpEx Costs - UF/MF Pretreatment ($/1000 gallons) =	$0.63	$0.63	$0.65	$0.71	$0.76

OpEx - Energy Cost Sensitivity Analysis

Membrane Life 5 yrs; Baseline Operating Pressure; $0.07/kWh; 189 km^3/d; 5% Cost of Capital

Energy Cost:	$ 0.04/kWh	$ 0.07/kWh	$ 0.10/kWh	$ 0.20/kWh
OpEx on Capital - Conventional Pretreatment 25-yr depreciation; 5% cost of capital	$0.25	$0.25	$0.25	$0.25
OpEx on Capital - UF/MF Pretreatment 25-yr depreciation; 5% cost of capital	$0.25	$0.25	$0.25	$0.25
Conventional Pretreatment	$0.06	$0.07	$0.08	$0.12
Chemicals ($/m^3)	$0.04	$0.04	$0.04	$0.04
Filters ($/m^3)	$0.00	$0.00	$0.00	$0.00
Energy ($/m^3)	$0.02	$0.03	$0.04	$0.08
UF/MF Pretreatment	$0.03	$0.03	$0.04	$0.06
Chemicals ($/m^3)	$0.01	$0.01	$0.01	$0.01
Membranes ($/m^3)	$0.01	$0.01	$0.01	$0.01
Energy ($/m^3)	$0.01	$0.01	$0.02	$0.04
First-Pass SW RO	$0.14	$0.22	$0.31	$0.58
Chemicals (antiscalant mostly, $/m^3)	$0.02	$0.02	$0.02	$0.02
Membranes (5-yr life, $/m^3)	$0.02	$0.02	$0.02	$0.02
Energy ($/m^3)	$0.11	$0.19	$0.27	$0.54
Second-Pass SW RO (optional)	$0.04	$0.05	$0.06	$0.10
Chemicals (NaOH mostly, $/m^3)	$0.01	$0.01	$0.01	$0.01
Membranes (5-yr life, $/m^3)	$0.01	$0.01	$0.01	$0.01
Energy ($/m^3)	$0.02	$0.03	$0.04	$0.08
Permeate Conditioning	$0.02	$0.02	$0.03	$0.04
Chemicals ($/m^3)	$0.02	$0.02	$0.02	$0.02
Energy ($/m^3)	$0.00	$0.01	$0.01	$0.02
Total Labor	$0.03	$0.03	$0.03	$0.03
Total Maintenance	$0.04	$0.04	$0.04	$0.04
Total OpEx Costs - Conventional Pretreatment ($/1000 gallons) =	$0.58	$0.69	$0.80	$1.16
Total OpEx Costs - UF/MF Pretreatment ($/1000 gallons) =	$0.55	$0.65	$0.76	$1.10

OpEx - Cost of Money Sensitivity Analysis

Membrane Life 5 yrs; Baseline Operating Pressure; \$0.07/kWh; 189 km^3/d			
Interest Rate:	3%	5%	7%
OpEx on Capital - Conventional Pretreatment 25-yr depreciation;	\$0.20	\$0.25	\$0.30
OpEx on Capital - UF/MF Pretreatment 25-yr depreciation; 5% cost of capital	\$0.21	\$0.25	\$0.31
Conventional Pretreatment	\$0.07	\$0.07	\$0.07
Chemicals (\$/m^3)	\$0.04	\$0.04	\$0.04
Filters (\$/m^3)	\$0.00	\$0.00	\$0.00
Energy (\$/m^3)	\$0.03	\$0.03	\$0.03
UF/MF Pretreatment	\$0.03	\$0.03	\$0.03
Chemicals (\$/m^3)	\$0.01	\$0.01	\$0.01
Membranes (\$/m^3)	\$0.01	\$0.01	\$0.01
Energy (\$/m^3)	\$0.01	\$0.01	\$0.01
First Pass SW RO	\$0.22	\$0.22	\$0.22
Chemicals (antiscalant mostly, \$/m^3)	\$0.02	\$0.02	\$0.02
Membranes (5-yr life, \$/m^3)	\$0.02	\$0.02	\$0.02
Energy (\$/m^3)	\$0.19	\$0.19	\$0.19
Second Pass SW RO (optional)	\$0.05	\$0.05	\$0.05
Chemicals (NaOH mostly, \$/m^3)	\$0.01	\$0.01	\$0.01
Membranes (5-yr life, \$/m^3)	\$0.01	\$0.01	\$0.01
Energy (\$/m^3)	\$0.03	\$0.03	\$0.03
Permeate Conditioning	\$0.02	\$0.02	\$0.02
Chemicals (\$/m^3)	\$0.02	\$0.02	\$0.02
Energy (\$/m^3)	\$0.01	\$0.01	\$0.01
Total Labor	\$0.03	\$0.03	\$0.03
Total Maintenance	\$0.04	\$0.04	\$0.04
Total OpEx Costs - Conventional Pretreatment (\$/1000 gallons) =	\$0.64	\$0.69	\$0.74
Total OpEx Costs - UF/MF Pretreatment (\$/1000 gallons) =	\$0.61	\$0.65	\$0.71

Appendix D

Estimating Unit Costs of Water Supply Options

The method for estimating water reliability benefits involves a two-step process. First, water managers define the level of reliability benefit they want to maintain or achieve. For example, they might want to ensure that enough water is available to meet demand in 39 out of 40 years, on average. Second, they compare options by adjusting average unit costs to get constant-reliability-benefit unit costs. The following example briefly illustrates the method (see Appendix D of Cooley et al. [2006] for the mathematical details).

Illustration of Constant-Reliability-Benefit Unit Costs

Suppose a community is served by supply from a local river with a normal distribution of hydrology.[1] Our example assumes the extractable yield in average years is 10,000 acre-feet (AF) and the standard deviation of annual flow is 1,000 AF. Low and high flows are increasingly rare as they get further from the average. The relative flatness of the bell is described by the standard deviation of the normal distribution. The larger the standard deviation as a percentage of the mean (this ratio is called the coefficient of variance), the flatter the bell, and the more variable is the annual flow available for human extractive purposes.

The average flow and the flow two standard deviations below the average are marked in Figure D-1. A property of the normal distribution is that in 2.5 percent of the years, flow will be less than the lower of these two marks. In our illustration, the flow two standard deviations

[1] The normal distribution is used for convenience. Hydrologic phenomena are usually better described by other distributions (e.g., log-normal, Pearson Type III, etc.).

below the mean is 8,000 acre-feet per year (AFY). Flow available for human use will be lower than the lower mark (8,000 AFY) in only 1 out of every 40 years over a long period of time.

Now let us consider demand. The demand numbers in our illustration are conveniently chosen to match some of the numbers in the description of supply, above. Any other numbers could be assumed, but they would make the illustration harder to follow. Assume that current drought-year demand (labeled D_E in Figure D-1)[2] is at the lower tick mark. Then the community served by this water system will experience a water shortage only 1 year out of 40. As defined above, this is a reliability level of 97.5 percent.

Suppose drought-year demand is projected to grow by 2,000 AF over the next decade.[3] As drought-year demand grows, reliability will decrease in the sense that the likelihood of a water shortage will increase from 1 in 40 to 1 in 2. That is, the reliability level would fall from

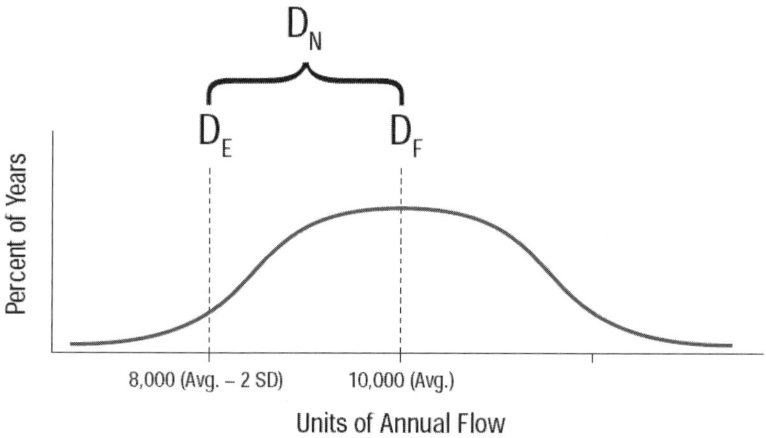

FIGURE D-1. Normal distribution of annual hydrologic flows.
SOURCE: Cooley et al. (2006).

---

[2] We define drought-year demand as the demand that would exist when flow is at a point chosen by the planner on the horizontal axis of Figure D-1—in this case, demand when flow is at the lower tick mark. Note that drought-year demand will often be higher than average-year demand because outdoor water use will increase when rainfall is below average or temperature is above average.

[3] A water demand projection is based on many factors, such as projected growth in population and employment in the service area.

97.5 percent to 50 percent, because enough water would be extractable in only half the years. Water managers may decide this is unacceptable and choose to maintain the current level of reliability at 97.5 percent. In this case, the amount of physical water (or water-use efficiency) required to satisfy growth in drought-year demand is the difference between future drought-year demand (D_F) and existing drought-year demand (D_E). This has been labeled D_N in Figure D-1, and in our example is 2,000 AF. If a supply option were to provide exactly this amount in every year, the planner should procure D_N of new supply. Water from advanced treatment processes (e.g., desalinated seawater or recycled wastewater) has this characteristic if treatment facilities are designed with enough redundancy to prevent downtime other than for regularly scheduled maintenance.[4]

But if the water supply option is variable from year to year, the planner must procure enough of it to have D_N available 39 out of 40 years, or reliability will decline. For example, when the chosen option is a surface water source, the amount available in an average year must be greater than D_N in order to ensure D_N is available in a dry year.

The amount of water supply greater than D_N that has to be purchased from the new water source depends on two factors: the new source's standard deviation of annual yield and the correlation of annual yield with the existing supply. The higher the new source's standard deviation of annual yield, the more water that needs to be procured from the new source to ensure adequate water in a low-flow year. The lower the correlations of annual yield between the new source and the existing source, the less of the new source will be required, on average, to ensure D_N is available in a dry year.

What this means is that comparing unit costs for options based on the average amount of water each option will deliver leaves out an important piece of the economic picture. For illustration purposes suppose that advanced treatment of impaired water, a new surface water supply, and outdoor conservation all have an average unit cost of \$600/AF. Ignoring reliability impacts, there is no financial difference between these sources.

But suppose further that the new surface water supply has a similar pattern of wet and dry years to the old surface water supply but is more variable. Then ensuring the 2,000 AF of new supply that will be needed in a drought year requires that the new source be sized to deliver more than 2,000 AF of water each average year, just as the old source was

---

[4] Some indoor water conservation measures may also have this characteristic of supplying exactly D_N every year if they are designed carefully.

capable of providing 10,000 AF on average but only 8,000 AF with the desired level of reliability. If the new surface water source has a coefficient of variance (the standard deviation over the mean) of 20 percent, the water planner will need to procure 3,333 AF in an average year to ensure 2,000 AF in the constant-reliability-benefit design year $(3{,}333 - (2 \times 0.2 \times 3{,}333) = 2{,}000)$. This in turn implies that each unit of water during drought will cost \$1,000/AF on a constant-reliability-benefit basis $(\$600/(1 - 2 \times 0.2) = \$1{,}000)$.[5] See Figure B-2 for an illustration of the average and constant-reliability benefit of surface water in this example.

If an outdoor water-conservation measure were to save more water during dry weather,[6] its constant-reliability-benefit unit cost would be less than the assumed \$600/AF. If it were perfectly countercorrelated with the current surface water source, and had a coefficient of variation of 10 percent, its constant-reliability unit cost would be $\$500/\text{AF} = (\$600/(1 + 2 \times 0.1))$. That is, ensuring 2,000 AF of water in a drought year would require outdoor conservation measures sized to deliver only 1,667 AF in an average year. The countercorrelation implies that, during a drought where flows in the current supply source are two standard deviations below its mean, outdoor conservation would save two standard deviations above its mean, which equals 2.0 when the mean is 1.667 and the standard deviation is 0.1667 (10 percent of the mean).

Figure D-2 summarizes the average unit costs and constant-reliability-benefit (drought-year) unit costs under these assumptions.

---

[5] Stated differently, the utility could pay 67 percent more *per average unit* of water from the advanced treatment facility (1000/600 = 1.67) compared to each average unit in the new surface water alternative—and provide the same economic benefit at the same cost to customers. Note that the premium is not in total, but per unit. The smaller advanced treatment facility is just as good as the larger surface water facility at reliably providing 2,000 AF, so a *per unit* premium is justified.

[6] For example, laser leveling, drip or microspray irrigation, scheduling improvements, ET controllers, and adjustments in sprinkler heads to improve distribution uniformity reduce the percent of applied water that percolates or evaporates. Since applied water must go up during drought, these measures will save more water during drought than during average or wet weather. Auto-rain shutoff devices, by contrast, save more water when it rains than when it is dry.

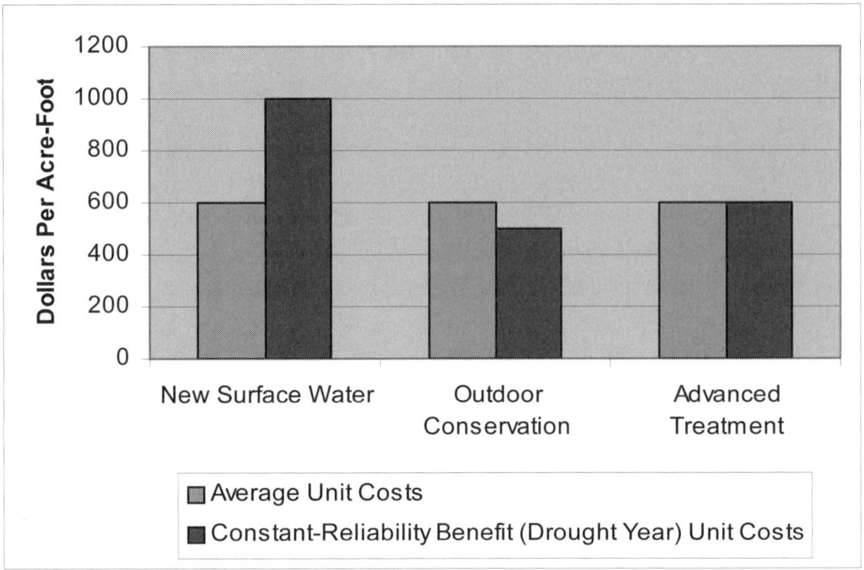

FIGURE D-2. Average and constant-reliability benefits of surface water alternatives, assuming equal average unit costs for each example. SOURCE: Cooley et al. (2006).

Accounting for variance and correlation between water sources—as is done for securities when managing a portfolio of financial assets—is clearly important. Water-supply planners who do not consider these factors might think options are similar in cost when they are in fact quite different once reliability benefits of the options are equalized. Worse yet, an apparently inexpensive source might turn out to be very expensive on a constant-reliability-benefit basis, or an apparently expensive source might turn out to have the lowest cost per acre-foot when reliability is considered.

SOURCE: Cooley et al. (2006).

Appendix E

Water Science and Technology Board

CLAIRE WELTY, Chair, University of Maryland, Baltimore County
JOAN G. EHRENFELD, Rutgers University, New Brunswick, New Jersey
GERALD E. GALLOWAY, University of Maryland, College Park
SIMON GONZALEZ, National Autonomous University of Mexico, Mexico City
CHARLES N. HAAS, Drexel University, Philadelphia, Pennsylvania
KENNETH R. HERD, Southwest Florida Water Management District, Brooksville, Florida
JAMES M. HUGHES, Emory University, Atlanta, Georgia
THEODORE L. HULLAR, University of California, Davis
KIMBERLY L. JONES, Howard University, Washington, DC
G. TRACY MEHAN, The Cadmus Group, Inc., Arlington, Virginia
DAVID H. MOREAU, University of North Carolina, Chapel Hill
THOMAS D. O'ROURKE, Cornell University, Ithaca, New York
DONALD I. SIEGEL, Syracuse University, Syracuse, New York
SOROOSH SOROOSHIAN, University of California, Irvine
HAME M. WATT, Independent Consultant, Washington, DC
JAMES L. WESCOAT, JR., University of Illinois at Urbana-Champaign

Staff

STEPHEN D. PARKER, Director
JEFFREY W. JACOBS, Scholar
LAUREN E. ALEXANDER, Senior Program Officer
LAURA J. EHLERS, Senior Program Officer
STEPHANIE E. JOHNSON, Senior Program Officer
WILLIAM S. LOGAN, Senior Program Officer

M. JEANNE AQUILINO, Financial and Administrative Associate
ELLEN A. DE GUZMAN, Research Associate
ANITA A. HALL, Senior Program Associate
DOROTHY K. WEIR, Senior Program Associate
MICHAEL J. STOEVER, Senior Program Assistant

Appendix F

Biographical Sketches for Committee on Advancing Desalination Technology

Amy K. Zander, Chair, is professor and director of the Interdisciplinary Engineering and Management program at Clarkson University, New York. Her research interests include drinking water treatment, treatment process design, membrane systems in environmental separations, life cycle assessment, and industrial ecology. Dr. Zander has received numerous awards for her research and teaching, including the 2003 Samuel Arnold Greeley Award from ASCE for the paper that makes the most valuable contribution to the environmental engineering profession, the 2000 AEESP/McGraw-Hill Award for Outstanding Teaching in Environmental Engineering and Science, and the 2001 Boeing Outstanding Educator Award. Prior to joining the faculty at Clarkson, she was a water quality specialist with the Texas Water Commission and an engineer with James M. Montgomery Consulting Engineers. Dr. Zander served on the NRC Committee on Small Water Supply Systems and the Committee on Assessment of Water Resources Research. She received her B.S. in biology and M.S. and Ph.D. degrees in civil engineering from the University of Minnesota.

Menachem Elimelech chairs the Department of Chemical Engineering and directs the Environmental Engineering Program of Yale University. He is also the Roberto C. Goizueta Professor of Environmental and Chemical Engineering. His research interests include fouling mechanisms of reverse osmosis and nanofiltration membranes, a novel forward osmosis desalination process, concentration polarization of interacting solute particles in crossflow membrane filtration, removal of natural hormones and pharmaceuticals by reverse osmosis and nanofiltration membranes, and membrane surface characteristics on membrane fouling. Dr. Elimelech is a member of the National Academy of Engineering and was awarded the 2005 Clark Prize for outstanding achievement in water science and technology. He received his B.S. and M.S. degrees from the Hebrew University of Jerusalem, and his Ph.D. in environmental engineering from the Johns Hopkins University.

David H. Furukawa is the president of Separation Consultants, Inc., a desalination consulting company. He has provided technical, management, and strategic assistance to institutions, communities, municipalities, nations, and private companies on desalination projects. Previously, he headed the saline water and demineralization section of the U.S. Bureau of Reclamation in the 1960s. He worked in various positions for UOP Fluid Systems, Boyle Engineering Corporation, Resources Conservation Corporation/Ionics, Inc., and FilmTech/Dow Chemical Company before heading Separation Consultants. Mr. Furukawa has served as president and director of the International Desalination Association and president of the American Desalting Association. Currently, he is chair of the Research Advisory Board for the National Water Research Institute and vice-moderator of the Research Advisory Council for the Middle East Desalination Research Center. He has authored more than 60 publications and is patented in the field. He received his B.S. in chemical engineering from the University of Colorado.

Peter Gleick is co-founder and president of the Pacific Institute for Studies in Development, Environment, and Security in Oakland, California, a nonpartisan policy research group addressing global environmental and development issues, especially in the area of freshwater resources. Dr. Gleick's research and writing address the hydrologic impacts of climate change, sustainable water use, desalination, privatization and globalization, and international conflicts over water resources. He is an internationally recognized water expert, and in 2003 he was named a MacArthur Fellow. He is a former member of the Water Science and Technology Board and was elected to the National Academy of Sciences in 2006. He received his B.S. in engineering and applied science from Yale University and his M.S. and Ph.D. degrees in energy and natural resources from the University of California, Berkeley.

Kenneth R. Herd is water supply program director of the resources projects department of the Southwest Florida Water Management

District. Formerly, he was the director of operations and facilities of Tampa Bay Water in Clearwater, Florida. He headed the operation and maintenance of a 250 million gallon per day water system, which has groundwater, surface water, and desalinated seawater components. He also oversaw facility construction, instrumentation and control, water quality, and other functions. From 2001-2008, he headed the Tampa Bay Seawater Desalination project, the largest seawater desalination plant in North America. Prior to this position, he managed engineering and project development efforts at Tampa Bay Water including the development of Tampa Bay Water's master water plan. He is a licensed professional engineer in Florida. He received his B.S. and M.S. degrees from the University of Kentucky in Lexington.

Kimberly L. Jones is associate professor of civil engineering at Howard University. Her research interests include methods for optimizing membrane processes for water treatment and biomedical applications, methods to reduce membrane fouling, mass transport, interfacial phenomenon, water and wastewater treatment plant design, and water quality. She is currently Deputy Director of the Keck Center for the Design of Nanoscale Materials for Molecular Recognition. Dr. Jones served on the NRC Committee to Review the Desalination and Water Purification Technology Roadmap and Committee on Environmental Decision Making: Principles and Criteria for Models. She joined the Water Science and Technology Board as a member in 2006. Dr. Jones received her B.S. and M.S. degrees in civil engineering from Howard University and the University of Illinois, respectively, and her Ph.D. in environmental engineering from the Johns Hopkins University.

Philip Rolchigo is the Vice President of Water Technologies at Pentair, Inc. Prior to this position he led the Global Separation Technologies section of General Electric, Water and Process Technologies. Formerly he was the chief technology officer and led the research and development division of Osmonics, Inc., a global manufacturer of high-technology water purification and separation technologies, which was later acquired by General Electric. He was also vice president of research, development, and engineering for Membrex, where he helped develop leading-edge membrane and filtration technologies. Dr. Rolchigo received his B.S. in chemical engineering from the University of Rochester in New York and his M.S. and Ph.D. degrees in chemical and biochemical engineering from the University of Pennsylvania.

Sandeep Sethi is a project manager and southeast region R&D lead at Carollo Engineers. His work experience spans over 15 years in desalination, membrane technology, and concentrate management projects. He has served as project manager, principal investigator, technical advisor, and lead process/project engineer on over 20 membrane technology projects that span diverse expertise including pilot, demonstration, and full-scale testing; predesign and design; process engineering; applied research; and planning and feasibility investigations. He has served as principal investigator and project manager on two AwwaRF projects on concentrate management and desalination/membrane technology. He has served on numerous professional committees and organizations focused on membrane technology that influence technology development, technical standards, and future regulations. He received his B.E. in civil engineering from Birla Institute of Technology and Science in Pilani, India, and his M.S. and Ph.D. degrees in environmental science and engineering from Rice University in Houston, Texas.

John Tonner is a senior consultant at Water Consultants International. He has over 18 years of experience in the design, engineering, and operations of both thermal and membrane desalination processes. Mr. Tonner has expertise in the technical basis of commercially viable desalination processes with experience from four continents and almost 50 countries and some of the world's largest and most technically advanced seawater desalination projects. He is a member of the board of directors of the International Desalination Association. He received his B.S. in mechanical engineering from the University of Paisley, Scotland.

Henry J. Vaux, Jr., is professor emeritus of resource economics at both the University of California, Berkley and Riverside. He is also associate vice president emeritus of the University of California system. He also previously served as director of California's Center for Water Resources. His principal research interests are the economics of water use, water quality, and water marketing. Prior to joining the University of California, he worked at the Office of Management and Budget and served on the staff of the National Water Commission. Dr. Vaux has served on the NRC committees on assessment of water resources research, western water management, and groundwater recharge, and, currently, sustainable underground storage of recoverable water. He was chair of the Water Science and Technology Board from 1994 to 2001. He received a Ph.D. in economics from the University of Michigan.

Judith S. Weis is a professor in the Department of Biological Sciences at Rutgers University. Her research has focused on estuarine ecology and ecotoxicology, mainly on stresses in the estuarine environment and their effects on organisms, populations, and communities. Particular areas of interest have been the effects of metal contaminants on growth, development, and behavior; development of tolerance to contaminants in populations living in contaminated areas; and effects of invasive marsh plant species on estuarine ecology, and on fate of metal contaminants. Much of her research has been focused on estuaries in the New York/New Jersey Harbor area. Dr. Weis served as an AAAS/American Society of Zoologists Congressional Science Fellow with the Senate Environment and Public Works Committee, and she was a Program Director at the National Science Foundation. She has been a member of the Marine Board of the NRC, and currently serves on the National Sea Grant Review Panel of the National Oceanic and Atmospheric Administration. She received her bachelor's degree from Cornell University and her M.S. and Ph.D. from New York University.

Warren Wood is the John Hannah Professor of Integrative Studies at the Department of Geological Science, Michigan State University. His research interests include groundwater and geomorphology of arid lands, hydrogeology, and groundwater geochemistry. Prior to joining the faculty of Michigan State University, he was as a research hydrologist at the U.S. Geological Survey until his retirement. His research experience includes the origin of salinity in the Sabkhas in the United Arab Emirates. He had also worked for many years in the High Plains area of Texas on a wide range of groundwater issues, including recharge and salinization and the origin of saline lakes. He received his B.S., M.S., and Ph.D. degrees in geology and hydrology from Michigan State University.

STAFF

Stephanie E. Johnson is a senior program officer with the Water Science and Technology Board. Since joining the NRC in 2002, she has served as study director for five committees, including the Review of the Desalination and Water Purification Roadmap and the Committee on Water System Security Research. She has also worked on NRC studies on contaminant source remediation, the disposal of coal combustion wastes, and Everglades restoration. Dr. Johnson received her B.A. from